Springer Texts in Education

Springer Texts in Education delivers high-quality instructional content for graduates and advanced graduates in all areas of Education and Educational Research. The textbook series is comprised of self-contained books with a broad and comprehensive coverage that are suitable for class as well as for individual self-study.

All texts are authored by established experts in their fields and offer a solid methodological background, accompanied by pedagogical materials to serve students such as practical examples, exercises, case studies etc. Textbooks published in the Springer Texts in Education series are addressed to graduate and advanced graduate students, but also to researchers as important resources for their education, knowledge and teaching. Please contact Natalie Rieborn at textbooks.education@springer.com or your regular editorial contact person for queries or to submit your book proposal.

More information about this series at http://www.springer.com/series/13812

Nicholas H. Wasserman
Timothy Fukawa-Connelly • Keith Weber
Juan Pablo Mejía Ramos • Stephen Abbott

Understanding Analysis and its Connections to Secondary Mathematics Teaching

Nicholas H. Wasserman
Teachers College
Columbia University
New York
NY, USA

Keith Weber
Graduate School of Education
Rutgers University
New Brunswick
NJ, USA

Stephen Abbott
Department of Mathematics
Middlebury College
Middlebury
VT, USA

Timothy Fukawa-Connelly
Temple University
Philadelphia
PA, USA

Juan Pablo Mejía Ramos
Graduate School of Education
Rutgers University
New Brunswick
NJ, USA

ISSN 2366-7672   ISSN 2366-7680   (electronic)
Springer Texts in Education
ISBN 978-3-030-89197-8   ISBN 978-3-030-89198-5   (eBook)
https://doi.org/10.1007/978-3-030-89198-5

This Springer imprint is published by the registered company Springer Nature Switzerland AG
The registered company address is: Gewerbestrasse 11, 6330 Cham, Switzerland

To all the teachers and students that helped us on the journey.

Preface

Getting certified to teach high school mathematics typically requires completing a course in real analysis. Yet most teachers point out real analysis content bears little resemblance to secondary mathematics and report it does not influence their teaching in any significant way. This textbook is our attempt to change the narrative. It is our belief that analysis can be a meaningful part of a teacher's mathematical education and preparation for teaching. The textbook is based on our efforts to identify ways that studying real analysis can provide future teachers with genuine opportunities to think about teaching secondary mathematics. This book is a *companion* text. It is meant to be used in conjunction with a more traditional analysis textbook. This choice allowed us to focus on how the ideas and experiences from a real analysis course can be applied to the secondary mathematics classroom without having to provide a full treatment of every topic in the course.

Real Analysis Content

The ideas in this companion text relate to a subset of topics typically discussed in an introductory analysis course. Chapter 1 lays out the foundational ideas for how teaching will be discussed in the form of six teaching principles. The next 12 chapters (Chaps. 2–13) correspond to 12 moments in a real analysis course where we make a point of connection to secondary teaching. Each chapter revolves around a particular definition, theorem, or proof from real analysis. The order of chapters follows the standard order in which the relevant analysis topics are typically encountered.

The following table outlines each chapter's relation to topics in *Understanding Analysis* (2nd ed.) by Stephen Abbott. (All forthcoming mentions of "Abbott" refer to this textbook.) Although we reference and use examples from Abbott's real analysis book throughout, our companion text is not specific to one textbook or another—the topics and themes would be found in other real analysis books as well. This table is a guide for when chapters in our textbook can be used during a real analysis course.

Companion Textbook	Abbott's Textbook	Analysis Content
Chapter 2 →	Sections 1.2–1.3	Real Numbers, Completeness
Chapter 3 →	Section 2.2	Sequence Convergence ($\varepsilon - N$)
Chapter 4 →	Section 2.3	Algebraic Limit Theorems for Sequences
Chapter 5 →	Sections 2.3–2.5	Monotone Convergence and Bolzano-Weierstrass Theorems
Chapter 6 →	Sections 4.2–4.3	Function Continuity ($\varepsilon - \delta$)
Chapter 7 →	Section 4.5	Intermediate Value Theorem
Chapter 8 →	Section 4.5	Inverse Functions, Continuity and Strict Monotonicity
Chapter 9 →	Section 5.2	The Derivative
Chapter 10 →	Sections 5.2–5.3	Derivative Rules
Chapter 11 →	Sections 6.5–6.6	Taylor's Formula and Taylor Polynomials
Chapter 12 →	Sections 7.1–7.4	The Riemann Integral
Chapter 13 →	Section 7.5	Fundamental Theorem of Calculus

As previously noted, our book presumes that the formal development of the real analysis content happens elsewhere. The chapters in this textbook represent *additional* readings and exercises for the definitions, theorems, and proofs listed above. The topics are discussed in a way intended to supplement and enrich—not replace—their treatment in a real analysis course. Subsequent chapters assume a familiarity with prior ideas and results. For instance, Chap. 7 in this companion text assumes you have encountered the material up to and including Abbott's Section 4.5 on the Intermediate Value Theorem (refer to table above). We do not provide a comprehensive treatment of the Intermediate Value Theorem. The agenda of this textbook is to explore how a formal approach to this result and others from an analysis course has relevance for the teaching of secondary mathematics.

Connections to Secondary Teaching

In every chapter, we highlight how the ideas in a real analysis course provide unexpected opportunities to think about teaching secondary mathematics in novel ways. Sometimes these opportunities for crossover relate to the *content* of secondary mathematics and other times they relate to ways that real analysis can be connected to the *process* of teaching. In this latter case, the analysis content becomes more peripheral to the pedagogical discussion, but the environment of the real analysis classroom still provides the primary catalyst and context for the conversation.

A *content connection* occurs when there is an overlap in the subject matter of real analysis with a topic explored in secondary mathematics. This starts with the real number system itself and extends through a surprising number of secondary school topics with roots in analysis—albeit, even still, we have not captured many possible mathematical connections between the two. For example, the concept of sequences from real analysis finds its way into the high school classroom in the form of decimal approximations for irrational numbers. We maintain that the improved understanding of irrational numbers that comes from elaborating on this connection to real analysis can lead to better secondary teaching. (Sequences and irrational numbers are the focus of Chap. 3.) The broad notion of area-preservation provides a different type of example. Integration and the properties of the Riemann integral studied in analysis supply a valuable framework for, and example of, the concept of area-preserving transformations studied in high school geometry. (This connection is the focus of Chap. 12.)

A *process connection* is an alignment or link that exists between the respective activities engaged in when studying analysis and when teaching secondary mathematics. These activities include making conjectures, creating definitions, solving problems, and so forth. Process connections are less about the topic being studied and more about the way we engage with the ideas. In chapters that explore process connections, the secondary content is typically *very different* than the analysis content. The bridge occurs in the type of reasoning that is employed. For example, creating proper definitions is a staple of real analysis as well as of many parts of secondary mathematics. In this light, introducing the $\varepsilon - \delta$ definition of continuity in real analysis becomes an opportunity to discuss the more general practice of defining as it relates to familiar geometric objects such as trapezoids and rhombuses. (This is the focus of Chap. 6.) With process connections, real analysis becomes a laboratory where we can reflect on the concrete experience of being a student to understand the mathematical nature of these activities and assess the analogous activities we oversee in the secondary mathematics classroom. These kinds of connections are opportunistic in the sense that the analysis content provides a context to discuss secondary teaching but in a way that does not necessitate the exact analysis content.

The Structure of Chapters

The structure of each chapter is based on the instructional model we used when we designed curricular materials for a real analysis course.[1] We referred to this approach as "building up from, and stepping down to, teaching practice."

Each chapter begins with a specific challenge that arises in teaching. We present this challenge as a short vignette from a secondary mathematics classroom, and readers are asked to reflect on how they might respond to that classroom situation. The pedagogical context then becomes the motivation for the mathematical content of the chapter. The chapters progress by expanding on the secondary mathematics topic from the vignette to uncover a connection to a particular component of real analysis, which in turn provides a new perspective on the original teaching

---

[1] Go to http://ultra.gse.rutgers.edu/ for more information about these materials.

situation. The final sections in each chapter highlight specific ways that teachers can incorporate the lessons learned from studying the real analysis content into their teaching. The discussions about pedagogy in these sections are aligned with the six teaching principles introduced in Chap. 1.

Each chapter concludes with a set of exercises that ask you to think more deeply about the ideas introduced in the chapter. Primarily focused on pedagogy, these exercises introduce related classroom situations and challenge you to think concretely about what you might do as a classroom teacher. There are also a few exercises that ask you to reflect on your personal experience learning real analysis and how it might inform your teaching. These are supported by a set of commentaries scattered throughout the book under the heading, "Turning the Tables: Reflecting on teaching from your learning in real analysis."

The chapters in this companion textbook are based on previously designed curricular materials. For readers familiar with those modules, this textbook is intended to be an additional resource that encapsulates the key ideas. Although similar in spirit, the two are not identical. Where this textbook goes further than the curricular materials is in clarifying—and elaborating on—the relationships between real analysis, secondary mathematics, and our set of teaching principles. There are several options for instructors planning to use both the curricular materials and this companion text in teaching a real analysis course: one is to assign the chapters in this textbook as readings *after* using the materials in class; another would be to assign parts of the chapters *prior* to using the materials in class and the rest of the chapter afterwards. In either case, the problems at the end of each chapter provide a wider range of potential assignments than those included in the curricular materials. We have not included sample solutions to any exercises within the text, but these can be made available to instructors.

The Audience

This textbook was developed with an audience of secondary mathematics teachers in mind and specifically for use while taking a course in real analysis. Each chapter provides a different answer to the question: How can ideas from real analysis be useful for teaching secondary mathematics? The chapters attempt to fill in the gaps between the ideas encountered in analysis and the pedagogical implications they have for teaching. But we hope the textbook might be used more broadly—not just with secondary teachers, and not just in real analysis courses.

Although our focus is on secondary teaching, any student taking real analysis can benefit from this companion book. The chapters provide an accessible complement to what happens in real analysis and would help any student gain a better understanding of the main ideas of this challenging course. Applying concepts in a new setting is a well-established practice for improving understanding (e.g., applying mathematical results to problems in business); bringing the ideas of real analysis to bear on high school teaching is an inventive example of this general principle, particularly because the setting is so relatable. We all have experiences we can draw on from our own secondary mathematics education, some more positive than others. Because mathematics classes are part of nearly everyone's journey, every student has something to contribute to these discussions regardless of whether they plan to become a teacher or not.

Although written to be a companion text for an analysis course, the text can certainly be used outside of this context. For example, it could be part of a capstone course for mathematics education majors where students are asked to make connections between their mathematical coursework and aspects of teaching. Alternately, individual chapters, or the specific pedagogical situations within them, might be the focus of a professional development workshop. We welcome and encourage other possible uses of this text, reminding potential readers that a familiarity with some results of introductory real analysis is a prerequisite for parts of every chapter.

Acknowledgments

The preparation of this companion textbook was based on work supported by a collaborative grant from the National Science Foundation (under DUE 1524739, DUE 1524681 and DUE 1524619). The focus of that project, *Upgrading Learning for Teachers in Real Analysis* (ULTRA), was on designing and testing curricular materials for use in a real analysis course—which then led to the idea for this companion text. Although there is strong alignment between the two, this textbook expands on the ideas from those curricular materials, compiling them into a meaningful resource for anyone interested in why, and how, learning real analysis can be germane to secondary mathematics teaching. More information about that project—including instructional materials, papers published from the project, and other supplementary files—can be found online at: http://ultra.gse.rutgers.edu/.

New York, NY, USA   Nicholas H. Wasserman

Philadelphia, PA, USA   Timothy Fukawa-Connelly

New Brunswick, NJ, USA   Keith Weber

New Brunswick, NJ, USA   Juan Pablo Mejía Ramos

Middlebury, VT, USA   Stephen Abbott
August 2021

Contents

1 **Six Teaching Principles** 1
   1.1 Theoretical Orientation of Our Teaching Principles 1
   1.2 Our Teaching Principles 2
        1.2.1 Teaching Principle 1 (TP.1): Acknowledge and
               Revisit Assumptions and Mathematical Constraints
               or Limitations 3
        1.2.2 Teaching Principle 2 (TP.2): Consider and Use
               Special Cases to Test and Illustrate Mathematical
               Ideas 3
        1.2.3 Teaching Principle 3 (TP.3): Expose Logic as
               Underpinning Mathematical Interpretation 4
        1.2.4 Teaching Principle 4 (TP.4): Use Simpler Objects
               to Understand More Complex Objects 5
        1.2.5 Teaching Principle 5 (TP.5): Avoid Giving Rules
               Without Accompanying Mathematical Explanations 5
        1.2.6 Teaching Principle 6 (TP.6): Seek Out and Give
               Multiple Explanations 6
   References 7

2 **Equivalent Real Numbers and Infinite Decimals** 9
   2.1 Statement of the Teaching Problem 9
   2.2 Connecting to Secondary Mathematics 10
        2.2.1 Problematizing Teaching and the Pedagogical
               Situation 10
        2.2.2 Equivalence Classes and Decimal Representations 11
   2.3 Connecting to Real Analysis 12
        2.3.1 An ε-Approach for Defining Equivalence of Real
               Numbers 14
        2.3.2 Implications for Real Numbers 15
   2.4 Connecting to Secondary Teaching 16
        2.4.1 Exploring Infinite Decimals with Students 17
        2.4.2 The Progression of Number Sets 18

Problems 20
Turning the Tables: Reflecting on TP.2 in *learning* analysis.............. 23
References.. 24

3   **Sequence Convergence and Irrational Decimal Approximations** 25
    3.1   Statement of the Teaching Problem 25
    3.2   Connecting to Secondary Mathematics 26
          3.2.1   Problematizing Teaching and the Pedagogical
                  Situation.. 26
          3.2.2   Approximations and Error-Bounds 27
    3.3   Connecting to Real Analysis....................................... 28
          3.3.1   Approximation Processes as Infinite Sequences 29
          3.3.2   Approximation Processes and Sequence Convergence ... 30
          3.3.3   Features of Approximation Processes 32
    3.4   Connecting to Secondary Teaching................................ 33
          3.4.1   Student-Centered Instruction............................. 33
          3.4.2   A General Heuristic Behind Approximations 35
    Problems 35
    Turning the Tables: Reflecting on TP.3 in *learning* analysis.............. 39
    References.. 40

4   **Algebraic Limit Theorems and Error Accumulation**................... 41
    4.1   Statement of the Teaching Problem 41
    4.2   Connecting to Secondary Mathematics 42
          4.2.1   Problematizing Teaching and the Pedagogical
                  Situation.. 42
          4.2.2   Approximation and Error Accumulation 43
    4.3   Connecting to Real Analysis....................................... 45
          4.3.1   The Algebraic Limit Theorem for sequences 45
          4.3.2   Implications for Error Accumulation 46
          4.3.3   Visualizing the Potential Error Inequality for Products... 48
    4.4   Connecting to Secondary Teaching................................ 49
          4.4.1   Applying Principles of Error Accumulation to
                  Design Problems... 49
          4.4.2   The Tip of the Iceberg 51
    Problems 52
    References.. 54

5   **Divergence Criteria and Logic in Communication** 55
    5.1   Statement of the Teaching Problem 55
    5.2   Connecting to Secondary Mathematics 56
          5.2.1   Problematizing Teaching and the Pedagogical
                  Situation.. 56
          5.2.2   Common Variants of Conditional Statements 59

5.3   Connecting to Real Analysis..   60
        5.3.1   Convergence Theorems   60
        5.3.2   Logical Implications About Divergence...................   63
5.4   Connecting to Secondary Teaching...............................   64
        5.4.1   Counterexamples   64
        5.4.2   Converses   66
        5.4.3   Grammatical Variation   67
Problems   68
References...   71

6   Continuity and Definitions...   73
6.1   Statement of the Teaching Problem   73
6.2   Connecting to Secondary Mathematics   75
        6.2.1   Problematizing Teaching and the Pedagogical
                 Situation...   75
        6.2.2   Trapezoids   76
        6.2.3   Isosceles Trapezoids......................................   77
6.3   Connecting to Real Analysis.....................................   78
        6.3.1   Considering Various Definitions of Continuity   79
        6.3.2   Choosing a Definition   83
6.4   Connecting to Secondary Teaching...............................   84
        6.4.1   Defining Isosceles Trapezoids   85
        6.4.2   The Relationship Between Definitions and Theorems   86
Problems   87
Turning the Tables: Reflecting on TP.4 in *learning* analysis...............   90
References...   91

7   The Intermediate Value Theorem and Implicit Assumptions   93
7.1   Statement of the Teaching Problem   93
7.2   Connecting to Secondary Mathematics   94
        7.2.1   Problematizing Teaching and the Pedagogical
                 Situation...   94
        7.2.2   Defining Function   95
7.3   Connecting to Real Analysis.....................................   96
        7.3.1   Differentiating Conditions in Statements.................   97
        7.3.2   Use of Conditions in Proofs..............................   99
7.4   Connecting to Secondary Teaching...............................   101
        7.4.1   Implicit Assumptions in the Classroom   102
        7.4.2   Implicitly Assuming the IVT in Secondary
                 Mathematics   103
Problems   105
Turning the Tables: Reflecting on TP.6 in *learning* analysis...............   108
References...   109

8   Continuity, Strict Monotonicity, Inverse Functions and Solving
      Equations 111
      8.1   Statement of the Teaching Problem 111
      8.2   Connecting to Secondary Mathematics 112
             8.2.1   Problematizing Teaching and the Pedagogical
                      Situation 112
             8.2.2   Solving Trigonometric Equations 113
             8.2.3   Inverse Functions 115
      8.3   Connecting to Real Analysis 116
             8.3.1   Inverse Functions and Strict Monotonicity 117
             8.3.2   Solving Equations 118
      8.4   Connecting to Secondary Teaching 119
             8.4.1   Strict Monotonicity as a Visual for Domain
                      Restrictions 119
             8.4.2   Inverses and Missing Solutions 120
             8.4.3   Inverses and Extraneous Solutions 122
      Problems 124
      References 126

9   Differentiability and the Secant Slope Function 127
      9.1   Statement of the Teaching Problem 127
      9.2   Connecting to Secondary Mathematics 128
             9.2.1   Problematizing Teaching and the Pedagogical
                      Situation 128
             9.2.2   Recognizing Computations as Singular Objects 129
      9.3   Connecting to Real Analysis 130
             9.3.1   The Secant Slope Function 130
             9.3.2   The Derivative as a Function 133
             9.3.3   Derivatives and Continuity 134
      9.4   Connecting to Secondary Teaching 136
             9.4.1   Navigating Disagreement in the Classroom 136
      Problems 138
      Turning the Tables: Reflecting on TP.1 in *learning* analysis 141
      References 142

10   Differentiation Rules and Attention to Scope 143
      10.1   Statement of the Teaching Problem 143
      10.2   Connecting to Secondary Mathematics 144
             10.2.1   Problematizing Teaching and the Pedagogical
                      Situation 144
             10.2.2   Attention to Scope 145
      10.3   Connecting to Real Analysis 146
             10.3.1   Scope in Theorems and Proofs 147
             10.3.2   Extending the Power Rule 148

10.4   Connecting to Secondary Teaching.................................   150
        10.4.1   Attention to Scope as a Pedagogical Practice   150
        10.4.2   Considering the Scope in Proofs Is Useful Too   152
Problems   153
Turning the Tables: Reflecting on TP.5 in *learning* analysis..............   157
References..   158

**11   Taylor Polynomials and Modeling the
      Complex with the Simple**   159
11.1   Statement of the Teaching Problem   159
11.2   Connecting to Secondary Mathematics   161
        11.2.1   Problematizing Teaching and the Pedagogical
                  Situation...   161
        11.2.2   Bounding and Approximating π with Regular n-gons ...   161
11.3   Connecting to Real Analysis...   164
        11.3.1   Taylor Polynomials   165
        11.3.2   Limits and Good Approximations   167
        11.3.3   The Real Reason for Analysis   169
11.4   Connecting to Secondary Teaching.................................   170
        11.4.1   Modeling the Complex with the Simple as a
                  Pedagogical Practice.......................................   170
        11.4.2   More with Circles...   171
Problems   173
References..   175

12   The Riemann Integral and Area-Preserving Transformations   177
12.1   Statement of the Teaching Problem   177
12.2   Connecting to Secondary Mathematics   179
        12.2.1   Problematizing Teaching and the Pedagogical
                  Situation...   179
        12.2.2   Area-Preserving Transformations and Cavalieri's
                  Principle...   180
12.3   Connecting to Real Analysis...   181
        12.3.1   Justification for Cavalieri's Principle via Integration   181
        12.3.2   Cavalieri's Principle and Integral Properties   183
12.4   Connecting to Secondary Teaching.................................   184
        12.4.1   Areas of Polygons   184
        12.4.2   Areas of Ellipses...   186
        12.4.3   Transformations That Are Not Continuous...............   187
Problems   188
References..   190

**13   The Fundamental Theorem of Calculus and Conceptual
      Explanation**   191
13.1   Statement of the Teaching Problem   191
13.2   Connecting to Secondary Mathematics   193

        13.2.1   Problematizing Teaching and the Pedagogical
                 Situation.. 193
        13.2.2   Rate of Change 193
        13.2.3   Area Conceptualized as in Cavalieri's (2D) Principle 194
   13.3 Connecting to Real Analysis... 194
        13.3.1   Instantaneous Rates of Area and $F'(c)$ in FTC(ii) 195
        13.3.2   Proof of FTC(ii) 197
   13.4 Connecting to Secondary Teaching................................. 198
        13.4.1   Conceptual Explanation 198
        13.4.2   Area and Conceptualizing Instantaneous Rates
                 of Change 199
   Problems 202
   References... 204

Afterword 205
   References... 209

Index 211

Six Teaching Principles

The Teaching Principles (TPs) introduced in this chapter guide how we link content from real analysis to situations in teaching secondary mathematics. They reflect what we consider to be positive characteristics of mathematics instruction. The TPs are not comprehensive. They do not attempt to capture all aspects of good teaching, which would be an impossible task. Rather, the TPs capture practices that we think are well-suited to being learned and discussed in a real analysis course. In part, this is because our TPs are premised on the belief that the way mathematicians do mathematics can be useful for how we think about teaching mathematics.

These principles are integral to this companion textbook. They ground our conversations about teaching in each chapter. Different aspects of a pedagogical situation, as well as possible next steps, are discussed in light of these TPs. They also serve as a way for you, as a student, to reflect on your own mathematical learning. Periodically, we incorporate sections entitled "Turning the Tables" to point out how these TPs may relate to your own learning of real analysis. The problems at the end of each chapter also invite you to think more deeply about the TPs by considering how they might be applied in other secondary teaching situations.

1.1 Theoretical Orientation of Our Teaching Principles

Broadly speaking, mathematics teaching is an amalgam of knowing *mathematics* and knowing *pedagogy*.[1] Our teaching principles similarly concern these two realms, but with regard to domains of practice instead of domains of knowledge.

Disciplinary practices are the kinds of activities practitioners of a discipline engage in when "doing." These include particular actions, habits of thinking,

---

[1] Shulman [7] described the intersection of these two domains of knowledge as *pedagogical content knowledge*.

© The Author(s), under exclusive license to Springer Nature Switzerland AG 2022
N. H. Wasserman et al., *Understanding Analysis and its Connections to Secondary Mathematics Teaching*, Springer Texts in Education,
https://doi.org/10.1007/978-3-030-89198-5_1

Fig. 1.1 Our TPs are
situated at the intersection of
mathematical and
pedagogical practices

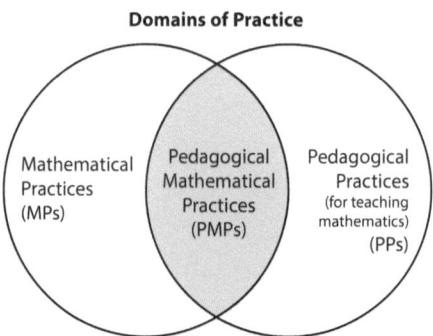

and lines of reasoning. So *mathematical practices* are the kinds of things that mathematicians do—activities such as conjecturing, generalizing, defining, formalizing, and proving (e.g., [1, 2, 6]). For *pedagogical practices*, which is really short for pedagogical practices for teaching mathematics, these are the kinds of things that (mathematics) teachers do—such as explaining, eliciting, facilitating, designing, and interpreting (e.g., [5, 8]). Our teaching principles are situated at the intersection of mathematical practices and pedagogical practices (Fig. 1.1)— what could be referred to as "pedagogical mathematical practices." The practices in this intersection manifest themselves differently in teachers than they do in mathematicians, but they strongly resonate with each other, sharing an affinity that mutually reinforces and informs one another. They capture how the practice of teaching is, in some sense, mathematical—how teaching meaningfully reflects the discipline being taught. Beginning with disciplinary practices is also a practical matter for us so that we can operationalize our pedagogical aims in relation to the mathematical coursework.

1.2   Our Teaching Principles

The six Teaching Principles (TPs) we describe below resonate with both mathematical practice and pedagogical practice. They are:

- TP.1. Acknowledge and revisit assumptions and mathematical constraints or limitations;
- TP.2. Consider and use special cases to test and illustrate mathematical ideas;
- TP.3. Expose logic as underpinning mathematical interpretation;
- TP.4. Use simpler objects to study more complex objects;
- TP.5. Avoid giving rules without accompanying mathematical explanations;
- TP.6. Seek out and give multiple explanations.

1.2.1   Teaching Principle 1 (TP.1): Acknowledge and Revisit Assumptions and Mathematical Constraints or Limitations

Many aspects of mathematics are about precision. Mathematicians aim to craft and prove statements in which the exact conditions are specified. Mathematical statements typically make all assumptions explicit and specify any constraints on what is or is not accomplished under those conditions. This mathematical practice is particularly evident in real analysis. Statements often begin with "Let $f : A \to \mathbb{R}$..." or "Let $f : [a, b] \to \mathbb{R}$ be continuous..." Subtle differences in the stated conditions are important (e.g., A vs $[a, b]$), but what is not stated is also important (e.g., continuity in the first statement). Indeed, specifying and revisiting assumptions such as continuity played an important role in the historical development of analysis as a field.

This mathematical practice is a productive practice for teaching mathematics as well. In mathematics, the expectation is to be explicit about assumptions or conditions and to be precise in every statement; in teaching, this may not always be desirable. It would be unproductive for an elementary teacher to stress that multiplication *on the natural numbers* is repeated addition. We would not expect elementary students to understand the nuance of what has been left unsaid by specifying "on the natural numbers" (i.e., but *not* on $\mathbb{Z}, \mathbb{Q}, \mathbb{R}$, or $\mathbb{C}$), given that $\mathbb{N}$ is the only set of numbers elementary students know! This does not mean that mathematics teaching should accept sloppiness or imprecision. For example, at the stage that students are learning to find the product of two rational numbers, it is important for teachers to acknowledge the limitations of "repeated addition" as the conceptual model for multiplication, and to help students transition to another model for multiplication to make sense of the operation in this new context (perhaps as "scaling"). Although it manifests itself differently between teacher and students than between mathematicians, acknowledging and revisiting assumptions and limitations is nonetheless an important pedagogical practice for teaching mathematics.

1.2.2   Teaching Principle 2 (TP.2): Consider and Use Special Cases to Test and Illustrate Mathematical Ideas

To probe the conditions and implications of a proposed theorem, mathematicians regularly consider special cases. This is incorporated in the process of crafting statements: looking at various discontinuous functions to determine whether continuity is a necessary condition or looking at the endpoints of an interval to determine whether a statement should be for $[a, b]$ or (a, b). It is also part of reconciling more general statements with the results of specific situations: validating that the law of cosines with a right angle yields the Pythagorean relationship or considering how limiting processes about secant lines imply statements about tangents. Whenever possible, mathematicians strive to identify a set of boundary cases to test or illustrate statements, either by crafting the special cases themselves or drawing on a well of known examples and counterexamples (e.g., the Dirichlet function in analysis; the Petersen graph in graph theory).

The use of special cases plays a similar role in pedagogy. Examples, counterexamples, and non-examples are important for learners because they make concrete what is initially abstract. Multiple theories of learning (e.g., [4]) suggest students assimilate concepts better by considering specific examples and non-examples that separate a phenomenon from its context and highlight its characteristic parts. Special cases can be used to help students differentiate features that are essential from those that are circumstantial or to clarify the boundary line between examples and non-examples. That said, the use of examples in mathematics teaching also merits caution. Mathematics is about relationships between abstract ideas characterized by specific definitions, and it is undesirable for students to overgeneralize from particular cases. For instance, when introducing the concept of function, showing students only examples of continuous functions or functions defined on $\mathbb{R}$ can erroneously suggest that all functions have these particular properties. Examples of discontinuous functions and functions between abstract sets are needed to more accurately convey the proper scope, or essence, of the definition. Effective teaching requires a strategic selection of examples to illuminate a definition, a statement, an argument, a proof, or a process—a mix that has sufficient breadth and variation to support rich concept development.

1.2.3   Teaching Principle 3 (TP.3): Expose Logic as Underpinning Mathematical Interpretation

Logic is part of the foundation of mathematics. Euclid's *Elements* is the archetypal example; from a basic set of axioms and definitions, over 400 propositions were deduced by applying the tenets of logic. Following this model, logic became the default language of mathematics whereby statements were crafted to clearly articulate the logical relationship between the ideas; e.g., 'If X, then Y' or 'X if and only if Y.' Eventually, mathematics turned its scrutiny inward so that logic itself became a free-standing area of mathematical study. Indeed, logic has come to permeate every facet of mathematics.

Teaching, meanwhile, necessarily takes place in the messy environment of social interactions and human communication. Although mathematical concepts and propositions are formally articulated, teachers and learners do not always convey or communicate about those ideas in strictly formal language. Mathematics teaching must give meaning—not just formal definition—to the mathematics being studied. Doing so necessarily means translating between formal ideas and everyday language. Words in everyday language can have a different meaning than they do in a mathematical setting, a phenomenon that can be especially challenging as it relates to the vocabulary of logic and deduction. The word "or" has a specific mathematical interpretation but in everyday language might be inclusive ($\vee$) or exclusive ($\oplus$). Likewise, the word "is" in one context might indicate a definition ($\iff$) and in another a one-way implication ($\implies$). Negotiating with the messiness and imprecision of language, which is part of the backdrop to teaching, requires a distinctive perspective when teaching mathematics. Pedagogical practice for teaching mathematics must acknowledge the necessity of informal classroom communication while consistently exposing the rigorous logical framework essential to mathematical discourse and interpretation.

1.2.4 Teaching Principle 4 (TP.4): Use Simpler Objects to Understand More Complex Objects

When confronted with a complex or unfamiliar object, mathematicians often proceed by modeling the complex object with a simpler, more familiar one. Archimedes' modeling of a circle with inscribed and circumscribed regular polygons in order to approximate π is an example from ancient times. There is a broad similarity between this approach and the problem-solving heuristic of replacing a difficult problem with a more accessible one. The mathematical practice described here is more specific, however. It refers to situations in which we gain insight into objects that are difficult to comprehend by modeling them with objects we already understand. Examples include using secant lines to understand tangent lines, using rational numbers to understand real numbers, using conical frustrums to understand spheres, using areas of rectangles to understand areas under curves, using line segments to approximate curves, using polynomials—via power series expansions—to understand more complicated functions, and so forth.

This mathematical practice is also useful as a pedagogical strategy. Learning is facilitated when new ideas are connected to old ideas that students know, and an organic way to achieve this in the classroom is by using familiar objects as the starting point to introduce unfamiliar ones. In this way, students become aware of this important mathematical practice at the same time as their instructors use it as an effective technique, and guide, for teaching. It can help teachers think about how to sequence topics so that new ideas build on results students already know. Providing opportunities for students to give justifications for new objects in terms they understand enables them to actively construct and organize their accumulating mathematical knowledge in a connected way.

1.2.5 Teaching Principle 5 (TP.5): Avoid Giving Rules Without Accompanying Mathematical Explanations

This TP is more general than the others, and its status as both a mathematical and a pedagogical practice more self-evident. In mathematics, proofs are paramount. To a mathematician, a proposition without a proof is an unfinished essay—an inert platitude relegated to second class status. In the *Elements*, Euclid refers to proofs as "demonstrations." They take the form of a set of instructions that, when actively followed, compel the reader to travel below the surface and experience the inevitable truth of the statement in question. The sense of illumination that a proof provides brings a theorem to life, upgrading it from a passive observation to a window of insight that leads to authentic understanding about some facet of the mathematical landscape. A mathematical statement without an accompanying explanation or demonstration is not really mathematics.

In terms of pedagogical practice, mathematics education unfortunately does not have a strong record of complying with this teaching principle. Most of us can recall being asked to memorize rules or algorithms in mathematics—ones we could neither justify nor, as a result, understand. The issue is that it can be difficult for learners to differentiate the multiple layers of understanding—not just *what* or *how* but *why*. Many people would say they "understand" the algorithm for multiplication of fractions: $\frac{a}{b} \cdot \frac{c}{d} = \frac{ac}{bd}$. Often, though, what they mean is they understand what to do and how to do it. But this is very different from understanding why! As a pedagogical practice, this principle requires that learners always be given the opportunity to understand the mathematical reasons for why specific rules, algorithms, procedures, and processes work. It is unfortunate that the "why" has not always been emphasized in mathematics instruction, as students are capable of understanding why most of the procedures they learn follow logically from the mathematical foundations that they know.

1.2.6   Teaching Principle 6 (TP.6): Seek Out and Give Multiple Explanations

Mathematicians regularly seek out different explanations for statements they already know to be true. To some, this mathematical practice might seem like wasted energy. Why spend time on something already known to be true? For mathematicians, each new explanation brings its own particular insight. The Pythagorean Theorem, for example, has been proven in over 350 different ways (cf., [3]). This practice is valued in mathematics because it leads to new understanding about, or appreciation of, the connections that exist between different domains of mathematics. These connections can serve as the impetus to pursue more general theories, but they also provide an opportunity to recognize the aesthetic nature of mathematics. With multiple options available, some proofs might be viewed as more satisfying because of their simplicity, their cleverness, their explanatory power, or their visualization. Indeed, this is an aspect of mathematics that is inherently beautiful: we can a approach a problem from a variety of perspectives and still arrive at the same conclusion!

Shifting our attention to pedagogical practice, we find that this principle is still deeply ingrained in teaching but for a slightly different reason. Using multiple explanatory approaches and perspectives is axiomatic in teaching. An underlying premise of differentiated instruction is that learners are different, so to engage all students in ways that are personally effective requires use of the multiplicity of available approaches. It is not just that some students will respond better to one approach than another, but that multiple approaches can help each individual learner glean important insights that are visible only from one particular approach or become evident as a commonality across several approaches. Engaging multiple explanations and approaches also offers students an opportunity to use their own voice. Just because a student approaches a problem differently than the teacher does not mean they are wrong. Differences can be used to affirm the individual as a creative thinker and as a launching point for others to engage in making sense of multiple perspectives—that is to say, thinking mathematically.

References

1. Cuoco, A., Goldenberg, E. P., & Mark, J. (1996). Habits of mind: An organizing principle for a mathematics curriculum. *Journal of Mathematical Behavior, 15*(4), 375–402.
2. Heid, M. K., & Wilson, P. S. (2015). *Mathematical understanding for secondary teaching: A framework and classroom-based situations*. Charlotte, NC: IAP.
3. Loomis, E. S. (1968). *The Pythagorean proposition* (2nd ed.). Reston, VA: NCTM.
4. Marton, F. (2014). *Necessary conditions of learning*. New York, NY: Routledge.
5. National Council of Teachers of Mathematics (NCTM). (2014). *Principles to actions: Ensuring mathematical success for all*. Reston, VA: NCTM.
6. Rasmussen, C., Zandieh, M., King, K., & Teppo, A. (2005). Advancing mathematical activity: A practice-oriented view of advanced mathematical thinking. *Mathematical Thinking and Learning, 7*(1), 51–73.
7. Shulman, L. (1986). Those who understand: Knowledge growth in teaching. *Educational Researcher, 15*(2), 4–14.
8. TeachingWorks. (2015). *High-leverage practices*. Retrieved (May 2018) from: http://www.teachingworks.org/work-of-teaching/high-leverage-practices

Equivalent Real Numbers and Infinite Decimals 2

2.1 Statement of the Teaching Problem

Mathematical ideas advance through progressively more powerful number systems:

$$\mathbb{N} \subseteq \mathbb{Z} \subseteq \mathbb{Q} \subseteq \mathbb{R} \subseteq \mathbb{C}.$$

In this chain, $\mathbb{N}$ represents the natural numbers, $\mathbb{Z}$ the integers, $\mathbb{Q}$ the rational numbers, $\mathbb{R}$ the real numbers, and $\mathbb{C}$ the complex numbers. With each step, students are faced with a heightened degree of abstraction. Natural numbers are concrete—they are easy to instantiate in the real world—whereas negative numbers, rational numbers, irrational numbers, and imaginary numbers are increasingly harder to conceptualize. One particular way students experience this escalation in abstraction is through the challenge of *representation*. As students move up to new number sets, not only do they need to learn different representation systems, such as representing real numbers as fractions or decimals, but one number might have multiple representations within a single system.

Consider the set of rational numbers, $\mathbb{Q}$. By definition, rational numbers are those that can be expressed as a quotient of integers; $\frac{a}{b}$ where $a, b \in \mathbb{Z}$ with $b \neq 0$. In contrast to the natural numbers, each rational number can be represented in multiple *different* ways. In fact, every one has an infinite class of equivalent representations. For instance, $\frac{1}{4}$ can be represented as $\frac{2}{8}$, $\frac{3}{12}$, $\frac{4}{16}$, and so on. Along with the loss of uniqueness that is inherent to fractional notation comes the added challenge that a rational number like $\frac{1}{4}$ can also be expressed as the decimal 0.25.

© The Author(s), under exclusive license to Springer Nature Switzerland AG 2022
N. H. Wasserman et al., *Understanding Analysis and its Connections to Secondary Mathematics Teaching*, Springer Texts in Education,
https://doi.org/10.1007/978-3-030-89198-5_2

Consider the following pedagogical situation:

> Ms. Schmidt asks her students to find the following sum:
>
> $$\frac{4}{5} + \frac{1}{4}$$
>
> One student, Mila, suggests that the answer is 1.05, having quickly converted to decimal representations to find the sum. "This is how I normally do it," she explains, "since fractions are harder to work with. There's only one way to write a decimal and you can add and subtract them in the normal way."

Mila's approach works well in this problem. Although she arrives at the sum in a way that the teacher did not expect, her process used meaningful mathematics and resulted in the correct answer. Indeed, the fact that there are often different ways to get to the same answer in mathematics is something to be celebrated! Yet in doing so Mila may have avoided the mathematical ideas intended by Ms. Schmidt, and some of the ideas Mila expresses have limitations that need to be addressed (TP.1). Figuring out how to respond to students when they approach problems in an unanticipated way is one of the challenges of teaching.

Before moving on, think about how you, as a teacher, might respond to the pedagogical situation just presented. What would you do next?

2.2   Connecting to Secondary Mathematics

2.2.1   Problematizing Teaching and the Pedagogical Situation

To unpack this pedagogical situation, we first problematize two responses to the student—ones that might be similar to some of your initial reactions.

One possible response is positive reinforcement. Mila has managed to solve a problem in a way that is potentially easier than other approaches and perhaps not expected by the teacher. Such a solution is commendable. Changing the numerical representation indicates she is being resourceful and not simply applying a memorized algorithm about how to add fractions with uncommon denominators. Furthermore, the solution may be a shortcut but it is not trivial. Switching from fractional to decimal form demonstrates a degree of mathematical competence. Nonetheless, there are some disadvantages to just offering congratulations. Specifically, the student has avoided developing proficiency with fractions, including the fundamental and necessary skill of writing fractions so they have common denominators. There are situations where converting fractions to decimals prior to performing an operation would be undesirable, or even impossible.

Another possible response is to ask the student to solve a different problem—one that reveals the limitations of the student's approach. Consider, for example, asking the student to find the sum, $\frac{1}{3} + \frac{1}{7}$. This introduces *infinite* decimal representations into the problem, significantly complicating matters. The unfamiliar nature of infinite decimals might be enough to encourage the student to return to fractional notation and attempt a solution using a common denominator. On the other hand, the student might be sufficiently happy with an approximate decimal answer like $0.333 + 0.143 \approx 0.476$. In the student's defense, approximations are often sufficient in applications (and on multiple choice exams!), and 0.476 is arguably more quantitatively accessible than the exact fractional answer of $\frac{10}{21}$. Still, there is a qualitative distinction between an approximate solution and a precise one that demands some attention. Not only might this distinction motivate the student to reconsider how they *operate* with fractions, it might prompt them to reconsider how they *conceptualize* rational numbers more generally.

2.2.2   Equivalence Classes and Decimal Representations

Moving from the natural numbers $\mathbb{N} = \{1, 2, 3, \ldots\}$ to the integers $\mathbb{Z} = \{\ldots - 3, -2, -1, 0, 1, 2, 3, \ldots\}$ to the rational numbers $\mathbb{Q} = \{\frac{a}{b} : a, b \in \mathbb{Z}, b \neq 0\}$ is motivated primarily by the desire to undertake more robust arithmetical operations. Addition and multiplication can be carried out in $\mathbb{N}$ without leaving the set, but we need $\mathbb{Z}$ if we want to properly define subtraction and $\mathbb{Q}$ if want to do division. The standard notation for numbers in these sets conveys the (accurate) impression that each successive set is constructed from the preceding ones; but something qualitatively different occurs in the description of $\mathbb{Q}$. Specifically, two distinct expressions like $\frac{1}{4}$ and $\frac{2}{8}$ can represent the same rational number. More formally, we say that two expressions $\frac{a}{b}$ and $\frac{c}{d}$ are equivalent if $ad = bc$. The collection of all such fractional expressions that share this property can then be bundled together into an *equivalence class*, and all members of this equivalence class are associated with a single rational number. The equivalence class for $\frac{1}{4}$ is the set $\left\{ \frac{1}{4}, \frac{-1}{-4}, \frac{2}{8}, \frac{-2}{-8}, \frac{3}{12}, \ldots \right\}$. While this set contains an infinite number of different fractional expressions, they all represent the same rational number.

The non-uniqueness of fractional representation is a familiar feature of the notation, and also a potential source of consternation for some students. In the pedagogical situation described above, the student's solution was to sidestep this issue altogether by changing from fractions to decimals. Although $\frac{1}{4}$, $\frac{2}{8}$, and $\frac{3}{12}$ are all different looking expressions, we write all of these with the same decimal, 0.25. At first glance, decimal representations do not appear to have the same equivalence issue that fractional ones do. Maybe—as the student claimed—there is only one decimal representation for any number. Stop and think for a minute: Do you think this is true?

One technical way to assert decimal expansions are non-unique is to point out that 0.25 can also be written as 0.250 or 0.2500. This trick of adding zeros can be applied to any terminating decimal, and a similar strategy can be used for rational numbers with repeating decimal representations; for example, representing $\frac{1}{6}$ as $0.1\overline{6}$ or $0.16\overline{6}$ or $0.166\overline{6}$. But this should feel unsatisfying, at least in mathematics.[1] While superficially different, $0.1\overline{6}$ and $0.16\overline{6}$ both describe the exact same infinite string of digits. They are certainly not as different as $\frac{1}{4}$, $\frac{2}{8}$, and $\frac{3}{12}$. To avoid this uninteresting technicality, let's agree to the following:

Statement A decimal representation implicitly refers to an *infinite* string of digits.

Because some numbers require an infinite decimal expansion, it's best to level the playing field and agree that every decimal expression contains an infinite string of digits. Thus 0.25, 0.250, and $0.25\overline{0}$ are all shorthand for the same infinite decimal $0.25000\ldots$. Likewise $0.16\overline{6}$ and $0.166\overline{6}$ are each ways to denote the infinite decimal $0.1666\ldots$.

With this convention, it may seem that decimal representations are indeed unique. In the next section we will explore this conjecture by looking more closely at examples like $0.25\overline{0}$ that have an infinite string of 0s in the tail. But here is the crux of the matter. Whereas using fractions to represent rational numbers emphasizes their algebraic properties, using decimals emphasizes their geometric properties. A decimal expansion is a description of a location, like an "address" on the real number line; when that address has an infinite number of directions to follow it is no longer precisely clear what it should mean. It makes intuitive sense, as long as you do not think about it too hard. Real analysis is the result of what happens when you do think about it too hard. Negotiating with infinite processes is the business of real analysis, and making rigorous sense of infinite decimal representations requires some better tools for determining whether two real numbers might in fact be equal, even if they are expressed using different decimal representations.

Could two different infinite addresses actually give directions to the same point on the number line?

2.3 Connecting to Real Analysis

The "real" in "real analysis" refers to the set of real numbers, $\mathbb{R}$, but what exactly are these numbers? The extension of the natural numbers $\mathbb{N}$ to the integers $\mathbb{Z}$ is relatively tangible in the sense that constructing a model of the bigger set—$\mathbb{Z}$ in this case—using the smaller set $\mathbb{N}$ as raw material is relatively intuitive. The same is true of the extension of $\mathbb{Z}$ up to $\mathbb{Q}$. To construct the rational numbers we take all possible quotients of integers.

The extension of $\mathbb{Q}$ to $\mathbb{R}$ is more conceptually challenging. When the ancient Greeks realized that certain geometric lengths like $\sqrt{2}$ and $\sqrt[3]{5}$ could not be

---

[1] In other fields, such as science, 0.25 and 0.250 are used to indicate a difference in measurement accuracy.

expressed as a ratio of integers—i.e., were irrational—their response was to prioritize geometry over arithmetic. Some 2000 years later, the 19th century efforts to firm up the logical foundations of calculus finally reached the point where a proper construction of $\mathbb{R}$ from $\mathbb{Q}$ was required, and several rigorous models were proposed. The details of these constructions will take us too far afield, but it's not too inaccurate to say that $\mathbb{R}$ is the result of filling in the gaps of $\mathbb{Q}$. Wherever $\mathbb{Q}$ has a hole, a new "irrational" number is defined and inserted into the number line to plug up the gap. Thus $\mathbb{R}$ is the disjoint union of the familiar rational numbers $\mathbb{Q}$ with these newly minted irrational numbers $\mathbb{I} = \mathbb{R} - \mathbb{Q}$.

This brings us to the issue of how to represent an arbitrary real number. Every rational number has a complete (and finite) description in the form $\frac{a}{b}$ where a and b are integers. A select few irrational numbers, which have acquired some degree of importance, have their own special notation, such as π, $\sqrt{2}$ and ϕ. But the common language used to describe both rational and irrational numbers is decimal notation. Every real number, rational or irrational, can be represented as a decimal.

Through the standard division algorithm for dividing b into a, a given rational number $\frac{a}{b}$ can be converted into a decimal. For rational numbers, this process results in a decimal expansion of a very specific form: it either terminates or begins cycling through a fixed periodic pattern. And those decimals that terminate are simply a particular kind of fixed periodic pattern—one with an infinite string of 0s at the end. Thus, rational numbers are precisely the real numbers with decimal expansions that are eventually periodic. Irrational numbers, then, are those with non-repeating decimal expansions. When we write an expression like $\pi = 3.141592\ldots$ or $\sqrt{2} = 1.4142135\ldots$, we have to acknowledge the insufficiency of these descriptions. The decimal representations keep going, and although there are algorithms for finding each successive digit, there is no tidy way to express the entirety of these expansions. Such is the nature of infinity and such is the reason for real analysis.

A preliminary task of real analysis is to verify our intuition that every real number can be represented as an infinite decimal and, conversely, that every infinite decimal describes a well-defined real number. This is a significant exercise. Each digit in a decimal expansion specifies the location of a number with progressively greater accuracy. The property of $\mathbb{R}$ that guarantees there really is at least one real number at that location is *completeness* in the form of the Nested Interval Property (cf., Theorem 1.4.1 from Abbott [1]). The property of $\mathbb{R}$ that guarantees there is at most one number at that location is the *Archimedean Property* (cf., Theorem 1.4.2 from Abbott). (Both the Nested Interval Property and the Archimedean Property can be derived from the standard Axiom of Completeness; cf., Abbott, p. 15.) Leaving the important details to a real analysis course, what emerges is confirmation that infinite decimal expansions can indeed be used to represent each and every real number.

While this is comforting, the thorny issue that remains is whether decimal representations are unique. Just as $\frac{1}{4}$ and $\frac{2}{8}$ are equivalent descriptions of the same rational number, can two distinct decimal expressions be equivalent in the same way?

2.3.1   An ε-Approach for Defining Equivalence of Real Numbers

The confirmation that real numbers correspond to decimal expansions, and vice-versa, brings with it the added benefit of employing geometric intuition to understand properties of $\mathbb{R}$. Given $a, b \in \mathbb{R}$, the expression $|a - b|$ provides a notion of the distance between a and b on the real number line. One trait for any reasonable notion of distance—defined on any collection of objects—is that *two objects should be equal if and only if the distance between them is equal to zero.* Now this statement may appear to be so obvious that it seems useless, but we can employ it to obtain a surprisingly helpful criterion for when two real numbers are the same (cf., Theorem 1.2.6 in Abbott):

Theorem Two real numbers a and b are equal if and only if for every real number $\varepsilon > 0$ it follows that $|a - b| < \varepsilon$.

Why is this criterion—that for every real number $\varepsilon > 0$, $|a - b| < \varepsilon$—logically equivalent to asserting that the distance between a and b is equal to 0? Before reading on, consider how you might prove this.

The proof amounts to considering two possibilities for the distance $|a-b|$. By the definition of the absolute value function, the distance between a and b will either be greater than 0 or equal to 0; that is, (i) $|a - b| > 0$, or (ii) $|a - b| = 0$. The strategy of the proof is to rule out (i) as a possibility so that the only option left is (ii).

Proof The first implication to prove is: $a = b \implies \forall \varepsilon > 0, |a - b| < \varepsilon$. This direction is sensible. If two real numbers a and b are equal (which is the condition), then the distance between them is 0 according to our definition—they occupy the same position on the number line. Because $|a - b| = 0$, it must be less than any positive ε, and we are finished.

The second implication is more interesting: $\forall \varepsilon > 0, |a - b| < \varepsilon \implies a = b$. Supposing the two real numbers are such that $|a - b| < \varepsilon$ for any positive ε (the condition), we have that every positive distance is too big to be the distance between them. So, the distance between a and b is smaller than 0.1, smaller than 0.01, smaller than 0.001, etc. Having ruled out the possibility that the distance between them is greater than 0, we conclude that the distance between them is precisely equal to 0, and a and b must be the same number.   □

This theorem gives us a new way of thinking about when two real numbers might be equal—one that is particularly useful for interpreting infinite decimal representations.[2]

---

[2] Again, these decimal expansions—and the geometric intuition that comes with them—rely on a version of completeness that includes the Archimedean Property.

2.3.2 Implications for Real Numbers

Two real numbers are equal precisely when the distance between them is zero, and the distance between a and b in $\mathbb{R}$ is equal to $|a - b|$. But how is the operation of subtraction carried out on two infinite decimals?

As a simple example, the standard algorithm for subtraction can be used to determine the precise distance between 0.25 and 0.249. Because $|0.25 - 0.249| = 0.001$, which is greater than 0, we can confirm what we've long believed—that the real numbers represented by 0.25 and 0.249 (or 0.25000... and 0.249000... using infinite decimal representations) are indeed different numbers. But how far apart exactly are, for example, $\pi \approx 3.141\ldots$ and $\phi \approx 1.618\ldots$? The first step in the standard algorithm is to subtract the two digits furthest to the right. But this does not make sense with infinite decimals. What digit is furthest to the right in π? Or ϕ? If we try starting from the left instead we run into trouble because we can never be sure about the need to borrow. In our $\pi - \phi$ example, if we subtract starting on the left, we have to borrow in the second step! Now, π and ϕ are of course different real numbers, which we could show using inequalities[3], but the point is that infinite decimals are not always amenable to being subtracted and this can muddy the water around deciding whether two real numbers are the same.

To make this concrete, consider the two infinite decimals $a = 0.250000\ldots$ and $b = 0.249999\ldots$. What is the distance between these two numbers? The subtraction algorithm loses meaning in this case, so how else might we determine $|a-b|$? Better yet, what can we determine *about* this distance? As a starting point, the distance cannot be 0.001—it must be smaller than this because $0.24\overline{9}$ is closer to 0.25 than 0.249 (which is 0.001 away). By a similar argument, $|a - b|$ is smaller than 0.0001 and also smaller than 0.00001. In fact, even though there seems to be no way to find the exact distance between these two real numbers by algorithmically computing $a - b$, we can rule out the possibility that $|a - b|$ is a positive number. Whatever positive number $\varepsilon > 0$ is proposed, we can argue $|a - b| < \varepsilon$. (Stop reading and sketch out a justification for this last statement.)

Based on the theorem, the only option left to conclude is that, perhaps counter-intuitively, $0.25\overline{0}$ and $0.24\overline{9}$ occupy the same position on the number line. The challenge with an infinite decimal representation is that if we think about $0.24\overline{9}$ as an infinite progression of numbers—as a process of "getting closer and closer to" a number—we have begun in the wrong spot. Instead, we should start by thinking of the infinite decimal $0.24\overline{9}$ as representing *one* number and occupying *one* position on the number line. The goal then is to figure out where. But this is the easy part. Once we recognize the infinite decimal $0.24\overline{9}$ as a properly defined real number, there is only one choice for its location—$0.24\overline{9}$ must be another name for the real number better known as 0.25.

The moral of this example is that decimal representations are not unique. In addition to illustrating normative approaches in an analysis course (i.e., formal

---

[3] As an example, we could show $\phi < 1.7 < 3.1 < \pi$, meaning $\pi - \phi > 3.1 - 1.7$, or $|\pi - \phi| > 1.4$.

Table 2.1 Infinite decimal representations for several rational and irrational numbers

Real number		
Rational number		**Irrational number**
Terminating decimal	Non-terminating rational	Non-terminating irrational
$\frac{1}{4} = 0.25\overline{0}$ or $0.24\overline{9}$	$\frac{1}{3} = 0.\overline{3}$	$\pi = 3.1415\ldots$
$\frac{7}{8} = 0.875\overline{0}$ or $0.874\overline{9}$	$\frac{5}{11} = 0.\overline{45}$	$1 + \sqrt{2} = 2.4142\ldots$
$\frac{2}{1} = 2.\overline{0}$ or $1.\overline{9}$	$\frac{4119}{9990} = 0.4\overline{123}$	$\phi^2 = 2.6180\ldots$

definitions and arguments involving ε), this conclusion is an important part of understanding the set of real numbers. While it is customary and largely appropriate to conflate the real numbers with the collection of all infinite decimal representations, there is a set of examples where different decimal expansions describe the same real number. That said, this curious phenomenon is more the exception than the rule. The previous example involves a terminating decimal, which turns out to be a necessary and sufficient ingredient for non-uniqueness.

Claim For any real number that can be expressed as a terminating decimal (which therefore makes it a rational number), there are precisely two infinite decimal representations—one of them ending in a string of 9s and the other in a string of 0s. Every other real number (which could be a rational with a non-terminating decimal or an irrational number) has a unique infinite decimal representation.

Table 2.1 summarizes these conclusions. We note the fact that the infinite decimals that are equal (in the terminating decimals column) are those ending in 0s or 9s, which is particular to using *base-10* numbers; Problem 2.10 at the end of the chapter asks you to think further about this.

2.4   Connecting to Secondary Teaching

In the initial teaching situation, the student's strategy of converting fractions to decimals to find the sum was clever, although it likely avoided the mathematics intended by the teacher. This in itself may not be problematic, but the student's approach has some limitations as well. Fractions with decimal expansions that do not terminate create arithmetic challenges, and decimal representations do not completely avoid the issue of equivalent representations that fractions present. Leaving these issues unaddressed would run contrary to teaching principle TP.1. Ideas from real analysis provide insight into some of the mathematical challenges that arise with decimal notation—particularly, because decimal notation necessarily requires us to consider the infinite.

2.4.1 Exploring Infinite Decimals with Students

One of the keys in responding to the student in the teaching situation is to think more generally about whether the student's strategy of switching from fractions to decimals would work in every case. The student's comment about there being only "one way to write a decimal" also needs to be tested. Doing so—as we just did—necessitates thinking about situations in which the approach might not work, or the comment not hold up. This work on the part of the teacher involves identifying assumptions and limitations. As TP.1 suggests, it is insufficient to simply identify such assumptions or limitations—the teacher must go further to explicitly acknowledge and revisit them with students.

One way to push back on the student's assumptions is to construct new problems that reveal their limitations. Finding examples that illustrate particular mathematical ideas is aligned with TP.2. The examples should problematize the student's approach for operating with rational numbers, as well as address some of the difficulties conceptualizing decimals themselves. You might choose examples to do each separately or examples that do both at the same time. Imagine the following scenario:

> In response, Ms. Schmidt asks Mila and the class to first try using decimals to add the following rational numbers, and then try adding them as fractions:
>
> $$\frac{1}{3} + \frac{2}{3}$$
>
> $$\frac{1}{30} + \frac{1}{15}$$
>
> $$\frac{5}{44} + \frac{3}{22}$$

Before reading on, make a list of what issues involving representations and arithmetic operations arise in each example.

In the first problem, students are likely familiar with these fractions as decimal representations: $\frac{1}{3} = 0.3333\ldots$ and $\frac{2}{3} = 0.6666\ldots$. Such an example provides an opportunity to think about the challenges of adding infinite decimals. Similar to the subtraction algorithm, the addition algorithm also goes from right to left, which is problematic for infinite decimals. In this particular example, however, it is relatively straightforward to visualize adding each column to obtain an infinite string of 9s. In this example, the student's contention that it is always easier to add decimal representations is turned on its head— here, the computation is messier with decimals! Another challenge to the student and the class arises from adding the fractions. Clearly $\frac{1}{3} + \frac{2}{3} = 1$, which forces students to consider their previous answer of $0.\overline{9}$ and the potentially disorienting conclusion that it must therefore be equal to 1. Now, it is also true that calculators will round $\frac{2}{3}$ to something like 0.66666667, and so students might argue that the decimals do not *really* add up to $0.\overline{9}$, they add up to 1 (as desired).

The decimal expansion of the fractions in the second example, $\frac{1}{30}$ and $\frac{1}{15}$, are likely less familiar. A calculator might help students write them as $0.03333\ldots$ and $0.06666\ldots$, which puts us back in essentially the same position as the previous example, with the sum being $0.09999\ldots$. Working with the fractional representations, we find a common denominator and compute $\frac{1}{30} + \frac{2}{30} = \frac{3}{30} = \frac{1}{10}$. Students would recognize this answer as the decimal 0.1, prompting the conclusion that 0.1 is equivalent to $0.0\overline{9}$. Providing multiple instances of this phenomenon reinforces the reality of non-unique decimal representations, but, as before, students may object by rounding the 'last' 6 into a 7 to avoid the dissonant idea that $0.09999\ldots = 0.1$.

The last example also uses fractions with unfamiliar decimal representations, and it generates a more complicated addition task. A calculator yields $\frac{5}{44}$ as $0.11363636\ldots$, and $\frac{3}{22}$ as $0.13636363\ldots$. The addition feels more challenging due to the alternating 3s and 6s, and the first few place-value sums being different than the rest. (Determining the difference between these two infinite decimals is even more challenging—you are asked to do this in Problem 2.3.) Still, the result of addition is relatively easy to imagine: $0.249999\ldots$. If instead we work with the fraction representations, find a common denominator, add, and then reduce, the result is $\frac{1}{4}$. Similar to the first two problems, this sum draws attention to the challenges of doing arithmetic with infinite decimals and the idea that two infinite decimals can be equivalent. In this example, the rounding issue is less likely to occur; rounding up to 0.113637 feels unlikely because the next decimal is a 3, which would round down to 0.113636. As this example illustrates, we can create other examples of two rational numbers whose sum results in an infinite tail of 9s by making sure the aligned columns in the repeating parts sum to 9—e.g., $0.11\overline{756} + 0.13\overline{243}$ (see Problem 2.2).

Asking students to compute examples like these prods them to consider the challenges of operating with rational numbers in their decimal form and explore issues about equivalent representations. As in these examples, there are times when decimals (not fractions) can be harder to work with! Indeed, why decimals like 0.25 and $0.24\overline{9}$ are equivalent can be more challenging to explain than fractions that are equivalent like $\frac{1}{4} = \frac{2}{8} = \frac{3}{12}$.

2.4.2   The Progression of Number Sets

The nested chain of number systems

$$\mathbb{N} \subseteq \mathbb{Z} \subseteq \mathbb{Q} \subseteq \mathbb{R} \subseteq \mathbb{C}$$

is a robust example of creating more complex objects out of simpler ones. Because each set is defined in terms of the previous one, we come to understand each more complicated number system in terms of the more primitive ones that come before it. This idea is connected to TP.4. In the present context we are exploring TP.4 as it applies in mathematics rather than teaching, but we typically mirror this kind of progression in secondary school mathematics as well. Each extension results in a number system with additional capabilities, but as we have seen there are some trade-offs as we move up the hierarchy. As new qualities are gained, others are lost. Some of these trade-offs are related to notational complexity. A recurring theme of this chapter is that in the step from $\mathbb{Z}$ to $\mathbb{Q}$ we lose uniqueness of representations, and from $\mathbb{Q}$ to $\mathbb{R}$ we encounter the delicate subtleties implicit in the infinite nature of decimal notation.

Beyond notation, the number systems themselves exhibit an interesting give-and-take of algebraic, order, and set-theoretic properties. In terms of algebra, $\mathbb{N}$ is closed under addition and multiplication, $\mathbb{Z}$ allows for subtraction, and $\mathbb{Q}$ makes it possible to properly define both subtraction and division. Moving to $\mathbb{R}$ allows for some new operations like square roots of positive numbers, and the complex numbers $\mathbb{C}$ can accommodate square roots of any number. As their algebraic dexterity increases, the sets become more crowded. The sets $\mathbb{N}$ and $\mathbb{Z}$ are discrete—each element has a unique successor, or "next largest" element in the ordering. This property is lost in the step up to $\mathbb{Q}$, which is still ordered but not in this discrete way. Given any two rational numbers a and b, the rational number $\frac{a+b}{2}$ sits in between them, as do infinitely more. This shows that the elements of $\mathbb{Q}$ are densely nestled together with no intervals of empty space. In every interval on the number line it is always possible to find rational numbers that are arbitrarily close together, but $\mathbb{Q}$ is still permeated by holes. The property of completeness that defines the step up to $\mathbb{R}$ fills in these holes, but it comes with a host of other implications. One of the most profound is that $\mathbb{R}$ is no longer a so-called "countable" set. While $\mathbb{N}$, $\mathbb{Z}$ and $\mathbb{Q}$ are infinite, there is a rigorous way to articulate that the infinity characterizing the size of $\mathbb{R}$ is of a distinct and higher magnitude. Although we have not met the complex numbers in this chapter, a price for moving from $\mathbb{R}$ up to $\mathbb{C}$ is that a meaningful ordering is no longer possible.

Taken together, as each successive number system is constructed, we see that it inherits some of the properties of its predecessor but also sacrifices others in pursuit of some other form of added dexterity. As the numbers change, the operations on them change, and the properties of each set need to be re-examined. When students transition from one number system to another, it is a good idea to address the possibility that properties they understood from a previous set may no longer apply: How do we need to think about arithmetic differently when we change to a new number set? Do the same conceptions of number, or operation, hold for these new objects? Do the same procedures work? How are these procedures dependent on the representation system being used? As students wrestle with different number systems and different notations for them, we need to remind them what is gained, and what is lost, in each case.

Problems

2.1 Write all possible infinite decimal representations for the following real numbers: (i) 0.8; (ii) 0.142; (iii) $\frac{15}{8}$; (iv) $\frac{9}{11}$; (v) $\pi/4$

2.2 Show that the sum of $0.11\overline{756}$ and $0.13\overline{243}$ yields an infinite tail of 9s. Determine fractions for these decimals, and show that the sum of the fractions is one-fourth. Find another pair of decimals whose sum would result in $0.249999\ldots$. Determine fractions for these decimals, and show that the sum of the fractions is one-fourth.

2.3 Use the infinite decimal representations of $\frac{3}{22} = 0.13\overline{63}$ and $\frac{5}{44} = 0.11\overline{36}$ to determine the difference. Compare your infinite decimal answer with what you would find the difference to be as a fraction.

2.4 Integers are 'signed' numbers. They afford the ability to differentiate numbers by adding a sign, $-$ or $+$. One result is they give (at least) two different ways to express a positive number. The representation '$+4$' would be one way to express the number 4. (i) What would be another way to express 4 using 'signed' numbers? [Hint: how might you express the additive inverse of '-4'?] (ii) Describe how this second representation of a positive number poses problems for students, and how you as a teacher might help address the problem.

2.5 Complex numbers are part of secondary mathematics. They are often written in the form $z = a + bi$ (with $a, b \in \mathbb{R}$) and plotted as the point (a, b) on the plane ($\mathbb{R}^2$). But complex numbers, and points on the plane, can also be referenced using a central angle, θ (rotation around $(0, 0)$ from the positive x-axis), and a radius, r (signed distance from $(0, 0)$)—like polar coordinates. In this way, complex numbers are written as $z = r(\cos\theta + i \sin\theta)$. If we allow our angle to be $\theta \in [0, 2\pi)$, and our radius r to be a real number (positive or negative), then there is an equivalence class on $\mathbb{C}$. What is the one other way you could express the complex number, $z = 3(\cos\frac{\pi}{4} + i \sin\frac{\pi}{4})$?

2.6 A student is trying to understand the idea that $0.999\ldots$ is equal to 1, and $0.24999\ldots$ is equal to 0.25, and so forth. In trying to generalize, the students asks: "So, then is $0.24777\ldots$ equal to 0.248?" (i) How would you respond to the student? (ii) Use the theorem in the real analysis section to justify whether these two numbers are equal or not.

2.7 A teacher asks a class to convert $\frac{8}{9}$ into a decimal. One student uses his calculator and says that the calculator has given 0.8888888889. The teacher responds, "Well, that's close, but there's actually never a 9, it's a bunch of 8s. The calculator is just rounding at the end." The student replies, "Well if it was 0.8888888888, then if I were to add $\frac{1}{9}$, which the calculator says is 0.1111111111, it would give me 0.9999999999. But it should be 1, that's why the calculator put the 9 at the end so that it would add to 1.0000000000." (i) Describe the mathematical ideas about real numbers that are being discussed. (ii) Provide a description for how you would respond to the student.

2.8 A pre-calculus teacher is trying to explain the limit of the following function: $\lim_{x \to \infty} \left(1 - \frac{1}{x}\right)$. In particular, that the the limit is *equal to*, and not just *close to*, 1. Describe how the idea of $\lim_{x \to \infty} \left(1 - \frac{1}{x}\right) = 1$ is similar to the idea that $0.\overline{9} = 1$. What is different about the situations mathematically?

2.9 Consider the following two proofs that $0.999\ldots = 1$. Discuss: (i) what assumptions are being made about infinite decimals in each proof; and (ii) in what contexts, if any, you might find either of these proofs useful for your own teaching.

Proof 1	Proof 2
Let $x = 0.999\ldots$	$\underbrace{0.999\ldots9}_{n} = 1 - \left(\frac{1}{10}\right)^n$
$10x = 9.999\ldots$	For a decimal with an infinite amount
$10x - x = 9x = 9.999\ldots - 0.999\ldots = 9$	of terms, we can use a limit:
Since $9x = 9$, then $x = 1$. Therefore	$0.999\ldots = \lim_{n \to \infty}\left[1 - \left(\frac{1}{10}\right)^n\right]$
$x = 0.999\ldots$ and $x = 1$, so $0.999\ldots = 1$.	$= 1 - \lim_{n \to \infty}\left(\frac{1}{10}\right)^n = 1 - 0 = 1$.

2.10 In this chapter, we saw that $0.24\overline{9} = 0.25\overline{0}$, or, more generally, that an infinite string of '9's was equivalent to 'bumping up' to the next decimal digit with an infinite string of 0s. This is in fact specific to *base-10* numbers. Now, consider base-6 numbers, which express numbers only using the numerals 0, 1, 2, 3, 4, and 5. Fractional decimals are similarly interpreted in that base. For example, $0.13000\ldots$, in base-6, represents 1 sixth $(1 \cdot \frac{1}{6})$ plus 3 sixths-squared $(3 \cdot \frac{1}{6^2})$. (i) What other base-6 infinite decimal would be equivalent to $0.13000\ldots$? Think about the criteria given in the theorem in the real analysis section for when two real numbers are equal. (ii) Describe the two infinite decimal expressions that would be equivalent for a general base-*b* number system.

2.11 Abbott's Sect. 1.1 begins the first chapter with a discussion of there being no *rational number* whose square is 2; later, in Sect. 1.4, this idea is revisited with the real numbers—that there is a *real number* whose square is 2. As an example of TP.2, how would you describe the mathematical idea that this particular example serves to illustrate? Now, discuss this same idea in relation to TP.1. [If you would like to do some further reading, the July 2020 edition of *Mathematics Teacher Educator* is about teaching and "mathematical statements that expire."]

2.12 In Sect. 1.2, Abbott gives an intuitive idea about sets: "Intuitively speaking, a *set* is any collection of objects" (p. 5). Afterwards, on p. 7, Abbott writes:

> Admittedly, there is something imprecise about the definition of set presented at the beginning of this discussion. The defining sentence begins with the phrase "Intuitively speaking," which might seem an odd way to embark on a course of study that purportedly intends to supply a rigorous foundation for the theory of functions of a real variable. In some sense, however, this is unavoidable. Each repair of one level of the foundation reveals something below it in need of attention. The theory of sets has been subjected to intense scrutiny over the past century precisely because so much of modern mathematics rests on this foundation. But such a study is really only advisable once it is understood why our naive impression about the behavior of sets is insufficient. For the direction in which we are heading, this will not happen, although an indication of some potential pitfalls is given in Sect. 1.7.

Describe the teaching principle that you believe Abbott is illustrating in this paragraph. Then, describe any way Abbott's use of this teaching principle here helps you think more generally about how this principle might be implemented in teaching.

Turning the Tables

Reflecting on *teaching* from your *learning* in real analysis: TP.2

Although the primary agenda of this supplemental textbook is to connect the content of a course in real analysis to secondary mathematics teaching, it would be a missed opportunity to ignore the teaching of analysis as a case study. An especially rich source of insight into good teaching comes from our personal experiences as students. Because users of this book are most likely currently engaged as students, it makes sense to pause from time to time and explore how a student's perspective in a challenging course like real analysis might illuminate the six teaching principles at the core of this book. We do so primarily by thinking about teaching as it is evident in an analysis text.

TP.2 is about the use of specific cases. One particular type of specific case are boundary cases, which are designed to test—or showcase—the limits of a definition, theorem, procedure, or proof. We touched on TP.2 briefly in response to the pedagogical situation in this chapter. To develop this principle further, and to think about it in the context of your own real analysis learning, consider an example Abbott introduces in Sect. 1.2: the *Dirichlet function*,

$$g(x) = \begin{cases} 1 : x \in \mathbb{Q} \\ 0 : x \notin \mathbb{Q} \end{cases}$$

He introduces this example immediately after defining function. The intent is to introduce a boundary case that illustrates the "unruliness" of functions (one that was also historically important in mathematics for the same reason). What is particularly effective about this example is that the Dirichlet function is regularly referenced throughout the remainder of the textbook. The Dirichlet function is useful to have in one's "example space" because it challenges our expectations of what a function is; in Abbott's words, "examples such as this one will provide us with an invaluable testing ground for the many conjectures we encounter" (p. 8). By introducing this example, Abbott is indicating the Dirichlet function to be so qualitatively different from other functions that it should refine how we think about functions and how we expect them to behave. This kind of example is the epitome of TP.2 in the way it shapes future thinking and learning. Be on the lookout for other instances of this teaching principle in your own learning of real analysis.

References

1. Abbott, S. (2015). *Understanding analysis* (2nd ed.). New York, NY: Springer.

Sequence Convergence and Irrational Decimal Approximations

3.1 Statement of the Teaching Problem

Calculators are powerful tools. With the press of a few buttons, they can compute products such as $0.25 \cdot 55 = 13.75$, quotients such as $12/7 \approx 1.7142857$, and approximations of square roots such as $\sqrt{12} \approx 3.46410162$. Although students can likely interpret all of these outputs, a question that arises is about proper use of the calculator. Is the calculator being employed to offload tasks that students have the capability to accomplish themselves, or is it being used as a "black box" that gives students a result they could neither produce nor justify? In this regard, the result of computing the product, 13.75, and even the quotient, $1.\overline{714285}$, are more readily accessible than the square root approximation, 3.46410162.

Yet secondary students are expected to learn how to approximate irrational numbers with rational ones and to estimate their position on a number line. Moreover, they are asked to learn *processes* for approximating irrational numbers; to be able to explain how to get increasingly better approximations. In this light, calculators should be used primarily as "task assistants" and not black boxes, facilitating computations so that students can focus on the conceptual and more challenging aspects of the mathematics.[1]

Consider the following pedagogical situation:

---

[1] Buchberger [2] elaborated a similar principle for the didactic use of symbolic computation (SC) systems: in the 'white-box' phase when students are still learning a mathematical topic the pertinent parts of the SC system are not to be used, but in the 'black-box' phase when students have mastered a mathematical topic the pertinent parts of the SC system are not only to be encouraged but essential.

© The Author(s), under exclusive license to Springer Nature Switzerland AG 2022
N. H. Wasserman et al., *Understanding Analysis and its Connections to Secondary Mathematics Teaching*, Springer Texts in Education,
https://doi.org/10.1007/978-3-030-89198-5_3

> Mr. Lopez is explaining how to give an approximate location of irrational numbers on the number line:
>
> Let's say we want to approximate $\sqrt{7}$. First, we know that 2^2 is 4, and we know that 3^2 is 9. Second, since $2^2 < 7 < 3^2$, we know that $2 < \sqrt{7} < 3$—that is, the value of $\sqrt{7}$ must be between 2 and 3. A similar technique can be used to figure out where other square roots, like $\sqrt{12}$, fall on a number line.

Mr. Lopez's explanation provides some information about where $\sqrt{7}$ is located on the number line (somewhere between 2 and 3). Indeed, the Common Core State Standards for Mathematics (CCSSM) [3] includes a similar example, explaining that "$\sqrt{2}$ is between 1 and 2," and then "between 1.4 and 1.5." He has also given a reasonable justification for this location—something that is valued in teaching (TP.5). And Mr. Lopez's approach may even contain a blueprint for getting better approximations. But upon closer inspection, although he has given a *range* for the irrational number, a particular *approximation* has not been given; and although the example might suggest an idea about how to get a better approximation, there is nothing explicit about this in his explanation. Identifying how one might accomplish these objectives, leveraging important characteristics about these kinds of mathematical processes, is one of the challenges of teaching.

Before moving on, think about how you, as a teacher, might explain approximating an irrational number, and what you might do next.

3.2 Connecting to Secondary Mathematics

3.2.1 Problematizing Teaching and the Pedagogical Situation

To unpack the pedagogical situation, we first problematize two potential next steps.

One next step would be to have students use a calculator. With the current estimate for $\sqrt{7}$, a calculator could serve two functions. The first would be to verify that the output, 2.6457513, is in fact between 2 and 3; the second would be as a means to provide further approximations—such as $2.6 < \sqrt{7} < 2.7$. Used in this way, however, the calculator is computing a value that students themselves have no other way to compute. Such an approach goes against TP.5 because it gives students a procedure to compute a square root ('use a calculator'), without providing an explanation why.

Another next step would be for the teacher to give some further approximations for $\sqrt{7}$ by hand. For instance, a teacher might continue by looking at $2.5^2 = 6.25$, concluding that this additional information means $\sqrt{7}$ is between 2.5 and 3; then, by checking $2.6^2 = 6.76$, and $2.7^2 = 7.29$, and thus concluding that $\sqrt{7}$ is between 2.6 and 2.7. In this case, the teacher's next step provides a better approximation for

Fig. 3.1 The location of $\sqrt{7}$ is somewhere in the interval $(2, 3)$

$\sqrt{7}$ and also provides a model for how this process might continue. Although this accomplishes part of the goal, the methods used do not have a consistent structure. In the second step, the teacher tested the midpoint of 2 and 3; in the third step, the teacher checked consecutive tenths. Although both work, these do not provide a clear template for students to get increasingly better approximations. The method is *inconsistent*—it does not generalize into a predictable process for approximation.

Although both of these next steps help students obtain a better approximation, there still remain some questions about why such procedures are effective. Specifically, while we might be convinced that using such a process results in increasingly better approximations, nothing so far allows us to be certain. We do not have a principled justification that our process always gets better in terms of approximating the desired irrational number.

3.2.2 Approximations and Error-Bounds

Let's return to the teacher's statement that $\sqrt{7}$ is between 2 and 3. The teacher's statement provides correct bounds on what $\sqrt{7}$ could be, but, if we wanted to locate it approximately on a number line, the statement is not clear about *where* it should be placed. Does it go in the middle of 2 and 3? Closer to 2? Or to 3? To summarize, the teacher's statement is about upper and lower bounds, not an approximation.

Definition Let a be a real number. A number $c \in \mathbb{R}$ is a **lower bound** for a if $c \le a$, and $d \in \mathbb{R}$ is an **upper bound** for a if $a \le d$. A **bounding interval** for a is a finite interval, such as $[c, d]$, within which a is definitely contained.

Definition Let a be a real number. An **approximation** for a is a specific $a_{appr} \in \mathbb{R}$ used to estimate the value of a.

Figure 3.1 corresponds to the teacher's statement: the interval $(2, 3)$ is a bounding interval for $\sqrt{7}$—we know it to be located somewhere strictly within the interval (as depicted). But the figure, as with the teacher's statement, does not contain a specific real number that is the approximation.

Although any number between 2 and 3 could be used as an approximation, the simplest choice initially is to use one of the endpoints. For illustrative purposes, let's take 2 as a first approximation for $\sqrt{7}$. To discuss the quality of an approximation like $a_{appr} = 2$ for the theoretical value $a = \sqrt{7}$, it's useful to distinguish between actual error and potential error:

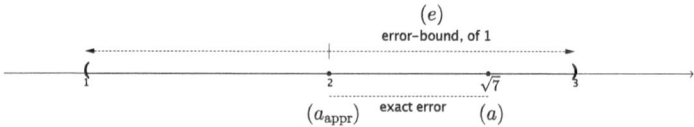

Fig. 3.2 Differentiating bound, error-bound, actual error and approximation

Definition The **actual error** between a theoretical real number a and an approximation a_{appr} is their difference on the number line, $|a - a_{appr}|$.

Definition The **potential error**, which we also refer to as an **error-bound**, for a theoretical real number a and an approximation a_{appr} is a positive real number e such that $(a_{appr} - e, a_{appr} + e)$ is a bounding interval for a. That is, e is a (strict) upper bound for the actual error, meaning $|a - a_{appr}| < e$.

Knowing that $\sqrt{7}$ is in the interval $(2, 3)$ means that our approximation $a_{appr} = 2$ comes with an implied error-bound of $e = 1$; the theoretical value $a = \sqrt{7}$ cannot be more than 1 unit away from our approximation. With error-bounds, there is a presumed symmetry to the bound—above *and* below the approximation. In our current example, this would yield the interval $(2 - 1, 2 + 1) = (1, 3)$ as a bounding interval for $\sqrt{7}$, but of course we previously determined $\sqrt{7}$ to be in the right half. The actual error is $\sqrt{7} - 2 \approx 0.64575$. This value is indeed less than 1. But, more to the point, if we knew the actual error then we would also know the theoretical value and have no need for an approximation in the first place! With approximations, an error-bound is generally more pertinent than the actual error. Figure 3.2 depicts possible locations for the theoretical number $a = \sqrt{7}$ with respect to the approximation $a_{appr} = 2$ and the error-bound $e = 1$; which is to say, a falls somewhere in the interval $(a_{appr} - e, a_{appr} + e)$.

As an an alternative to choosing 2 as our approximation, we could have used another number such as 2.5 (the midpoint) to approximate $\sqrt{7}$. In this case, if we kept $(2, 3)$ as our bounding interval, then we could say the error-bound is $e = 0.5$ since $(2.5 - 0.5, 2.5 + 0, 5) = (2, 3)$. Note that we are not providing the actual error; we are stating that the approximation's actual error is no worse than the error-bound e.

3.3   Connecting to Real Analysis

One aim of learning about approximating irrational numbers, according to the CCSSM, is to have a way to "continue on to get better approximations" [3]. The charge to "continue on" toward better approximations provides a point of connection to real analysis. If we had an algorithm or a process that allowed us to continue to generate improved approximations, then a running list of these successive approximations

$$\left(a_{appr_1}, a_{appr_2}, a_{appr_3}, \ldots \right)$$

generates a sequence. Sequences, which by definition contain an infinite number of terms, are a fundamental object of study in real analysis.

3.3.1 Approximation Processes as Infinite Sequences

Let's consider two possible processes for approximating $\sqrt{7}$.

The easiest to imagine is a sequence of truncated decimals:

$$(a_n) = (2, 2.6, 2.64, 2.645, \ldots) .$$

A teacher might explain these approximations in relation to $\sqrt{7}$ with the following observations: $2^2 < 7 < 3^2$; $2.6^2 < 7 < 2.7^2$; $2.64^2 < 7 < 2.65^2$; $2.645^2 < 7 < 2.646^2$; etc. First we find pairs of consecutive integers with squares that bound 7; then pairs of consecutive tenths with squares that bound 7; then consecutive hundredths; and so on. Choosing the lower bound as our approximation in each case generates the sequence (a_n). Each approximation in the sequence comes with an implied error-bound since the error is, at most, the difference between the upper and lower bounds. Table 3.1 summarizes this iterative process. The sequence (a_n) consists of the numbers found in the approximation column. Notably, the error-bounds have the consistent form, $\left(\frac{1}{10} \right)^{n-1}$.

If you had to determine the next entry in the sequence, what would you do? Try to produce the approximation for iteration 5 before moving on.

In this first approach, as you likely experienced, there is no particularly efficient way to determine the next bounding pair of real numbers. A systematic way to do so is to check all pairs, beginning at $(\ldots \underline{1}, \ldots \underline{2})$ and continuing on. For iteration 5, you stop when you discover $2.645\underline{7}^2 < 7 < 2.645\underline{8}^2$. Admittedly inefficient, the process is at least systematic and achieves the desired result.

A more efficient algorithm can be formulated using midpoints of intervals. Like our first approach, we start with 2 as the first estimate of $\sqrt{7}$, which we know to be in the interval $(2, 3)$. Now choose the midpoint, which in this case is 2.5, and ask whether $2^2 < 7 < 2.5^2$ or $2.5^2 < 7 < 3^2$. Here, we find $2.5^2 < 7 < 3^2$. As recorded in iteration 2 from Table 3.2, we can be certain $\sqrt{7}$ is within this new interval; the

Table 3.1 A sequence of truncated decimal approximations and error-bounds for $\sqrt{7}$

Iteration	Bounds	Approximation	Error-bound
1	(2,3)	2	1
2	(2.6,2.7)	2.6	0.1
3	(2.64,2.65)	2.64	0.01
4	(2.645,2.646)	2.645	0.001
n	(c_n, d_n)	c_n	$\left(\frac{1}{10} \right)^{n-1}$

Table 3.2 A sequence of midpoint approximations, and error-bounds, for $\sqrt{7}$

Iteration	Observed bound	Approximation	Error-bound
1	$(2, 3)$	2	1
2	$(2.5, 3)$	2.5	0.5
3	$(2.5, 2.75)$	2.75	0.25
4	$(2.625, 2.75)$	2.625	0.125
n	(c_n, d_n)	$\frac{1}{2}(c_{n-1} + d_{n-1})$	$\left(\frac{1}{2}\right)^{n-1}$

estimate 2.5 in that iteration is the tested midpoint—the midpoint of the *previous* iteration's observed bounding interval—which gives a new error-bound of 0.5. The next iteration begins by testing the midpoint of the interval $(2.5, 3)$, which is 2.75. After testing the square of 2.75, we arrive at $2.5^2 < 7 < 2.75^2$, which gives us a new bounding interval for $\sqrt{7}$. The error-bounds in each subsequent iteration are halved, which means there is a general form for the nth iteration. Table 3.2 provides details for these midpoint approximations, which produce the sequence:

$$(b_n) = (2, 2.5, 2.75, 2.625, \ldots).$$

In both examples, the algorithm can be repeated indefinitely. This means the approximation process produces a sequence, which has an infinite number of terms. Moreover, in both examples the error-bound is decreasing at each step.[2] Does this mean the processes are *actually* approximating $\sqrt{7}$? What more might we need to know besides the fact that the error-bounds get smaller?

3.3.2   Approximation Processes and Sequence Convergence

In analysis, we have a particular way of understanding whether a sequence of approximations is tending toward the desired value as the iterations increase. That idea is *arbitrary closeness*. This means that for an arbitrary error-bound ε, no matter how small, our approximations a_{appr} will at some point be within that level of closeness to a. This is the essence of the definition of convergence of a sequence.

Definition A sequence (a_n) **converges** to a real number a if, for all $\varepsilon > 0$, there exists an $N \in \mathbb{N}$ such that whenever $n \geq N$ it follows that $|a_n - a| < \varepsilon$.

The expression $|a_n - a|$ represents the actual error between a theoretical value a and a term in the sequence a_n. Note that a_n refers to the nth term in the sequence; for our purposes, this nth term can be understood as an approximation (e.g., $a_n =$

---

[2] Although we do point out a trade-off: while we tested fewer values in the midpoint process, the error-bounds do not get smaller as quickly as in the truncated decimal process. This means we need more iterations to achieve a similar level of precision.

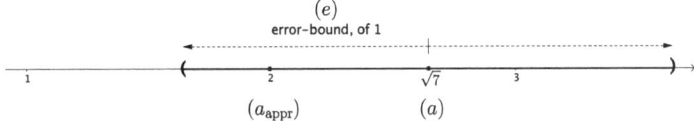

Fig. 3.3 Approximations falling within an ε-error-bound of a

a_{appr_n}). The definition specifies that the actual error be *no worse than* ε when $n \geq N$, meaning the approximation a_N and all subsequent approximations have ε as an error-bound.

In the convergence definition, the inequality $|a_n - a| < \varepsilon$ can be interpreted with the emphasis on either a or a_n. In the former case, the theoretical value a is at the center of the interval and the approximations a_n must be within the interval $(a - \varepsilon, a + \varepsilon)$. In the latter case, the approximation a_n is at the center and the theoretical value a falls within $(a_n - \varepsilon, a_n + \varepsilon)$. The former interpretation is the typical way to geometrically picture the definition of convergence (see Fig. 3.3), but the latter is more in the spirit of our discussions in this chapter. Both are viable ways to interpret $|a_n - a| < \varepsilon$.

The question at hand is whether the two processes for approximating $\sqrt{7}$—truncated decimal or midpoint—produce estimates that eventually become arbitrarily close to the desired value. In analysis terms, we are asking whether either or both of the sequences created from these processes converges to $\sqrt{7}$. Let's consider the truncated decimal process which produced the sequence $a_1 = 2$, $a_2 = 2.6$, $a_3 = 2.64$, $a_4 = 2.645$, etc. (You will look at the second midpoint process in Problem 3.2.) The approximation process required identifying consecutive numbers of a fixed number of decimal places with squares that bound 7 on either side. The nature of this algorithm means that $\left|a_1 - \sqrt{7}\right| < 1$; $\left|a_2 - \sqrt{7}\right| < \frac{1}{10}$; $\left|a_3 - \sqrt{7}\right| < \frac{1}{10^2}$; etc. In other words, the process guarantees the sequence (a_n) satisfies

$$\left|a_n - \sqrt{7}\right| < \frac{1}{10^{n-1}}$$

for each $n \in \mathbb{N}$.

Pause here and think about what the definition of convergence requires, and whether our sequence (a_n) meets it.

Claim The sequence (a_n) of truncated decimal approximations converges to $\sqrt{7}$.

Proof For some small error-bound, our goal is twofold: to show the truncating decimal process—the sequence—eventually produces an approximation within that error-bound of $\sqrt{7}$, and to show all subsequent terms in the "tail" of the sequence are also within that error-bound.

Let's first try a small error-bound of $0.000001 = 10^{-6}$. Looking back at Table 3.1, we can project that on iteration 7 (which is the term a_7 in the sequence) the error-bound will be equal to 10^{-6}. The next iteration, term a_8, will have an error-bound strictly less than this value. Hence, the process will produce an approximation within 10^{-6} of $\sqrt{7}$. Since the error-bounds are decreasing, all subsequent terms in the tail—i.e., after a_8—will have error-bounds less than 10^{-6} as well.

Now, let's carry out the same process with an arbitrary $\varepsilon > 0$. Because each a_n is within $\frac{1}{10^{n-1}}$ of $\sqrt{7}$, the aim is to find a natural number N such that $\frac{1}{10^{N-1}} < \varepsilon$. Ordering properties help substantiate the following: if $N > 1 - \log_{10} \varepsilon$, then $\log_{10} \varepsilon > 1 - N$, which means $\varepsilon > 10^{1-N}$, and so $\varepsilon > \frac{1}{10^{N-1}}$. This sequence of algebraic steps shows that choosing $N > 1 - \log_{10} \varepsilon$ does the trick. For such an N, the term a_N in the approximation sequence is within ε of $\sqrt{7}$. As before, we now argue that the error-bounds are decreasing. This happens because each successive error-bound is obtained by multiplying by $\frac{1}{10}$. Thus we can be confident that all terms a_n where $n \geq N$ will also be within ε of $\sqrt{7}$.

Since we were able to find this Nth term in response to an arbitrary error-bound $\varepsilon > 0$, we can do it for any value, no matter how small. This means the sequence of truncated decimal approximations becomes arbitrarily close, or converges, to $\sqrt{7}$.

$\square$

3.3.3 Features of Approximation Processes

What do proofs about sequence convergence reveal about our processes for approximating irrational numbers?

One important feature of our algorithms is that they are *iterative*—the nth approximation is used recursively to determine the $(n + 1)$th approximation. In the truncated decimal process, 2.6 was one of our approximations. This value was then the starting point for the next set of intervals: $(2.6\underline{0}, 2.6\underline{1})$, $(2.6\underline{1}, 2.6\underline{2})$, $(2.6\underline{2}, 2.6\underline{3})$, etc. Testing which of these intervals has endpoints whose squares bound 7 resulted in 2.6\underline{4} being chosen as the next approximation, which was then used to determine the following approximation of 2.64\underline{5}. Because each stage is used to determine the next, we implicitly have an *infinite* number of approximations, which is a required aspect of the definition of a sequence in the formal language of real analysis.

A second feature is *consistency*, by which we mean the approximating process applies the same computational procedure at each stage. In the truncated decimal process, this procedure might be described as: (i) using the previous approximation, add one additional decimal place and test each digit $(1-9)$ in that place to determine a new interval with squares that bound 7; (ii) take the lower bound of this new interval as the next approximation. Consistency is what facilitates communication; with it, we do not need an infinite number of instructions to communicate about an infinite process. More importantly, it is easier to think about and work with processes that are consistent. As a case in point, the consistent nature of the process allows us to *generalize* in algebraic terms how the nth term in the sequence relates to the theoretical value. In the truncated decimal process, each approximating interval had length equal to one-tenth the length of the previous one, which led directly to the relationship $\left|a_n - \sqrt{7}\right| < \frac{1}{10^{n-1}}$. Expressing each term's error-bound as $\frac{1}{10^{n-1}}$ then allowed us to work more concretely in the proof. The generalized, consistent expression was critical for identifying a sufficiently large N for a given ε, as well as for justifying that subsequent terms would remain within that same error-bound.

A third feature necessary for convergence is that the error-bounds of successive approximations become *arbitrarily close* to 0. At each iteration, progress is associated with the error-bounds becoming smaller, meaning our approximations are indeed getting closer to the theoretical value. There is, however, a subtle and important distinction between saying the error-bounds are decreasing and saying they are getting arbitrarily close to 0. It is this latter quality that means our process effectively approximates the theoretical value. In the truncated decimal process, the error-bounds are of the form $\frac{1}{10^{n-1}}$, and a sufficiently large n makes this expression as close to 0 as we would like.

In sum, these three features represent the kinds of properties that processes for approximation should entail.

3.4   Connecting to Secondary Teaching

In the initial teaching situation, the teacher provided some ideas about approximating an irrational number but not with enough detail to allow students to meaningfully engage in an approximation *process* (at least as described by CCSSM standards [3]). Use of a calculator to compute the square root would serve as a rule without any mathematical explanation, violating TP.5. Calculator issues aside, the teacher's explanation and alternative next steps did not meet the standards we have discussed for an approximating process. In the initial situation, the teacher's explanation was not iterative—the process stopped; in a proposed next step, the explanation was inconsistent—it used a mixture of methods to find the subsequent approximation.

3.4.1   Student-Centered Instruction

It is relatively easy to imagine elaborating on the teacher's initial example by further explaining how one might obtain increasingly accurate approximations. This might look like the truncated decimal process, the midpoint process, or some other one. Here, however, we discuss a different next step that we believe clarifies why having a sense of the three highlighted features—iterative, consistent, arbitrarily small error-bounds—is especially beneficial in a classroom context.

Although good explanations are an important part of teaching, some of the most powerful learning occurs when students are given a degree of *agency* to create and discover. The topic of approximation provides such an opportunity. Consider the following extension of the teaching situation:

Rather than having students move on to approximate $\sqrt{12}$, Mr. Lopez instead asks students to work in groups. Building on the fact that they now know $2 < \sqrt{7} < 3$, he asks the groups to do the following:

1. In the next 5–10 min, work with your group to find an even smaller interval—the smallest interval you are able to—that you know contains $\sqrt{7}$
2. Write out instructions about how your group found this smaller interval, which you will give to another group, and that will allow them to replicate your group's process.

By asking students to determine their own approximation processes—instead of telling them what to do—the teacher creates some new opportunities. First and foremost, students have to do some mathematical thinking with their peers which can deepen their own learning. Before moving on, think about what else might be gained by giving agency to students in the classroom.

Asking students to come up with their own approaches likely means there will be a variety of proposed algorithms. This might even happen within each group. Asking each group to write out instructions anticipates these internal differences and requires the students to consolidate their various ideas into a singular approach. These differences will also surface across groups. Although this can be challenging to navigate in the classroom, we regard it as an affordance. Having groups share their processes with each other exposes students to multiple approaches to improving their approximations, an instructional practice aligned with TP.6.

Another consequence of giving students agency is that some of their strategies will be unanticipated. Students do strange things. Sometimes their ideas will be exceptionally clever and sometimes they will be erroneous or lack proper reasoning. Most likely their proposals will fall somewhere between these two extremes. To be able to address unanticipated contributions in real-time, teachers need *general* ways to evaluate the strategies their students propose. Here is where the insights we gleaned about approximating processes from real analysis proofs about sequence convergence pay real dividends. By evaluating whether a proposed approximation process is iterative, consistent, and achieves arbitrarily small error-bounds, teachers can effectively engage with their students' contributions. These insights allow teachers to be more flexible in their instruction, equipping them to comment on a variety of approaches rather than just prescribing a particular one.

3.4.2   A General Heuristic Behind Approximations

There is a more general heuristic going on in our conversation about approximations that relates directly to TP.4: Using simpler objects to study more complex objects. Generating decimal approximations for an irrational number is an instance of this principle.

Irrational numbers are unwieldy and a bit mysterious. The original discovery that $\sqrt{2}$ was irrational meant some numbers could not be expressed as a ratio of integers and thus could not be represented in the notational systems in use at that time. (In ancient Greek, "irrational number" translates roughly as a "number without a name.") The base number systems we use today largely solve this problem, although a little mystery remains. Decimal expansions for irrational numbers are necessarily infinite and that infinite quality leads to conceptual challenges. How do we make proper sense of an infinite decimal expansion? One answer is to think of it as a sequence of rational approximations. One of our sequences for approximating $\sqrt{7}$ is

$$(2, 2.6, 2.64, 2.645, 2.6457, \ldots).$$

Each of the values in the sequence is a *rational* approximation for the *irrational* number $\sqrt{7}$. These simpler rational approximations become a productive and practical way to understand the more complex notion of an irrational number. Indeed, the most famous proof that the irrational numbers are not countable (a surprising result about there being different sizes of infinity) employs expressing irrational numbers as infinite decimal expansions!

Approximating irrational numbers with decimals employs the heuristic of using simpler objects to understand more complex ones. But TP.4 suggests something more—specifically that this principle should be used in *teaching* mathematics. When teaching students about irrational numbers, we can help them understand these new numbers by modeling them with the rational numbers they already know. As an example, students sometimes think $\sqrt{2} + \sqrt{5} = \sqrt{7}$. One way to show why this is incorrect is by using rational approximations. We know $\sqrt{2} \in (1.4, 1.5)$, and $\sqrt{5} \in (2.2, 2.3)$, so their sum must be between 3.6 and 3.8. But since $\sqrt{7} \in (2.6, 2.7)$, we can conclude $\sqrt{2} + \sqrt{5} \neq \sqrt{7}$. Drawing conclusions about how irrational numbers work based on rational approximations reinforces the principle articulated in TP.4 as productive not only in mathematics, but also in mathematics teaching.

Problems

3.1 Suppose you wanted an approximation of $\sqrt{7}$ that was within $0.000001 = 10^{-6}$. For the truncating decimal sequence (a_n), we found that this would be true on the 7th term (i.e., $N = 7$). Compare this with the midpoint sequence (b_n). For what value N would the midpoint sequence become within $\varepsilon = 10^{-6}$?

3.2 Consider the process used to create the sequence (b_n) of midpoint approximations for $\sqrt{7}$. (The first few values of the sequence are depicted in Table 3.2.) (i) Write out a description of the consistent 'procedure' used to generate this sequence; give it to someone else to see if they can follow your instructions for creating the approximations in the sequence. (ii) Use the definition of convergence to prove that a sequence defined according to this process would converge to $\sqrt{7}$.

3.3 In the truncated decimal approximation process, we used the lower bound as our approximation each time. Think about what would result if we had used an identical process (checking for consecutive "ones", "tenths", etc., whose squares bound 7), but rather than using the lower bound as our approximation each time, we had used the midpoint of the bounding interval. The table below gives the first two iterations. (i) Complete the rest of the table—make the error-bound of the approximation as small as you are able for each iteration. (ii) What is the error-bound for the nth iteration—the error-bound for each term a_n in the sequence.

Iteration	Observed bound	Approximation	Error-bound
1	(2,3)	2.5	
2	(2.6,2.7)	2.65	
3	(2.64,2.65)		
4	(2.645,2.646)		
n	(c_n, d_n)		

3.4 Suppose someone created a sequence approximating $\sqrt{7}$ using both upper- and lower-bounds, so that the sequence was: 2, 3, 2.6, 2.7, 2.64, 2.65, ... (i) Determine the error-bound for each term in the sequence. (ii) Use the definition of convergence to show this sequence also converges to $\sqrt{7}$. (Think about how, and why, the value of N in this example differs from the one for (a_n) in the chapter.)

3.5 Develop a sequence that approximates $\sqrt{12}$. (i) Carry out the iterative steps until your error-bound is no greater than 0.0001. Specify the approximation and the error-bound at each step. (ii) Justify your sequence will converge by showing (a) you can find a value of N such that the term a_N has an error-bound less than some arbitrary ε, and (b) the error-bounds in the 'tail' of the sequence are no greater than the error-bound in the identified step. (iii) Write a paragraph that explains your process in a way that you could tell eighth graders, as well as a short justification you could give them about why the process approximates $\sqrt{12}$.

3.6 Consider the following process to approximate $\sqrt{7}$. As from before, let's begin with observed bounds of $(c_1, d_1) = (2, 3)$. From the bounding interval, select a *random real number* a_1 in that interval which will be your approximation. Test whether $a_1^2 < 7$ or $7 < a_1^2$; if the former, the next iteration begins from the interval $(c_2, d_2) = (a_1, d_1)$, if the latter it begins from the interval $(c_2, d_2) = (c_1, a_1)$. Repeat this process. Create a table generating the few first approximations to get a feel for this process. Now, describe which, if any, of the three features of approximation processes this procedure lacks.

3.7 Three approximations for the rational number $\frac{12}{7}$ are 1.7, 1.71, 1.714. The long-division algorithm is a procedure that can be used to generate approximations for $\frac{12}{7}$; successive stages of this procedure generated each of the three approximations above, and it could be used create an infinite sequence: $(a_n) = 1.7, 1.71, 1.714, \ldots$. First, provide a written description of this procedure (which should describe one stage of an iterative process). Make sure you could use it to explain to a student how they could continue to get additional digits in the decimal approximation for $\frac{12}{7}$ and why that process makes sense for division. Second, discuss how the long-division algorithm exemplifies the three features of approximation processes—iterative, consistent, and error-bounds arbitrarily close to 0. (Note: "terminating" decimals such as 0.25 should be considered, as they were in Chap. 2, as having repeating 0s at the end, i.e., $0.25\overline{0}$.) Third, explain where, at each stage in the procedure, the expression $\left| a_n - \frac{12}{7} \right|$ is produced, and describe how you might use that to prove this process will produce a sequence whose limit is $\frac{12}{7}$. Lastly, the process "iterates" based on a "remainder"; and the possibilities for this remainder depend on the divisor (the number in the denominator). Use this to explain why every rational number must have a repeating chunk of digits in its infinite decimal approximation.

3.8 The real roots of a function, $f(x)$, are those real numbers that make the equation $f(x) = 0$ true. For any particular value, x_0, the trichotomy principle states: $f(x_0) < 0$; $f(x_0) = 0$; or $f(x_0) > 0$. We can use this idea to determine a root of a continuous function (since they maintain the Intermediate Value Property we discuss later in Chap. 7). We can do so by identifying increasingly small x-intervals that bound the y-values of the function around 0. Look at the graph of $f(x) = x^3 - 4x^2 - 11x + 2$ below.

$f(x) = x^3 - 4x^2 - 11x + 2$

(5.9, 3.24)

x_0

(5.8, −1.25)

We have zoomed in to one of the visible roots, x_0, and we have plotted two points: when $x = 5.8$, $f(x) = -1.25$, and when $x = 5.9$, $f(x) = 3.24$. These points suggest something about the zero: $x_0 \in (5.8, 5.9)$. Using this as a starting point, come up with a process for approximating this one root of f. Try to find increasingly small intervals in which the zero x_0 is contained. You might create a table to keep track of your process. The process for approximating a root using a graphing calculator typically involves providing a left- and a right-estimate; can you make sense of how a calculator might iteratively find an approximation for a zero of a function?

3.9 One topic of study in Algebra II or Pre-calculus is the study of polynomials and various theorems about their roots. The Location Theorem for Polynomials states: "for $a, b \in \mathbb{R}$, if the sign of $f(a)$ and $f(b)$ changes, then there exists a root of the polynomial between a and b." (Such an idea was likely helpful in Problem 3.8.) First, in consideration of TP.3, think about interpreting this statement using *logic*. Namely: (i) What does the conclusion mean about "there exists a root"?—i.e., draw several different graphs that would showcase different ways in which the function might meet the conditions of the conclusion; (ii) The statement is an implication of the form $p \implies q$; is p the *only* way for q to occur, or could it be that $p \iff q$?—i.e., justify the biconditional or draw at least one counterexample. Second, describe how you might use the approximation process described in Problem 3.8 to help students learn about the Location Theorem for Polynomials.

3.10 In introducing what Abbott states is "arguably the most important definition in the book" ([1], p. 43), he explicitly calls attention to two different, albeit equivalent, definitions for convergence. First, he gives the standard $\varepsilon - N$ analysis definition for sequence convergence; a paragraph or two later, he gives a topological version of the definition (in terms of ε-neighborhoods). Describe what teaching principle you believe is being illustrated here. Explain your reasoning.

Turning the Tables

Reflecting on *teaching* from your *learning* in real analysis: TP.3

As an opportunity to reflect further on mathematics teaching, we explore another way in which your own learning of real analysis might exemplify one of our teaching principles. Specifically, we look at this in relation to making sense of the $\varepsilon - N$ definition of sequence convergence.

TP.3 states that teachers should expose the logic underpinning mathematical interpretation. Formal mathematical logic is a useful tool for interpreting mathematical statements, and attending to logic can be productive in teaching and learning. Indeed, recasting common observations into logical language is an important part of mathematical and pedagogical practice because it highlights the role logic plays in mathematics. Logical clarity and precision, such as being explicit about whether statements are conditional ($\Rightarrow$) or biconditional ($\Leftrightarrow$) relationships, is evident across the entirety of an analysis course. Here, we discuss the quantifiers $\forall$ and $\exists$, which are central to the $\varepsilon - N$ definition of sequence convergence.

As a first example of attending to these logical aspects, Abbott discusses divergence with an example sequence, $\left(1, -\frac{1}{2}, \frac{1}{3}, -\frac{1}{4}, \frac{1}{5}, -\frac{1}{5}, \frac{1}{5}, -\frac{1}{5}, \ldots\right)$ (Example 2.2.8). The example is used to point out that while a sequence might be amenable to *some* values of ε (i.e., he shows ε could be $\frac{1}{2}$, so there is at least one), if it does not work *for all* ε values (a counterexample of $\varepsilon = \frac{1}{10}$ is given) the sequence does not converge. He also provides meaning around the existence of an N—that there be at least one N where a_N falls within the ε-neighborhood and so does a_n for every $n \geq N$. These nuances around logical quantifiers are particular to mathematics. In everyday communication, the word "some" can be ambiguous because the criterion for "how many" depends on the context. For mathematicians, the emphasis is typically on whether a property holds in (i) all cases, (ii) no cases, or (iii) at least one case. In logical terms, "some" refers to options (iii) and (i).

As another example of using logic to help students learn about mathematics, Abbott asks readers to consider what would happen if the order of the quantifiers in the $\varepsilon - N$ definition of convergence were reversed (Exercise 2.2.1): "A sequence (x_n) 'verconges' to x if *there exists* $\varepsilon > 0$ such that *for all* $N \in \mathbb{N}$ it is true that $n \geq N$ implies $|(x_n) - x| < \varepsilon$." The exercise asks learners to think about how switching quantifiers impacts the types of sequences being described. This forces students to grapple with the logical structure of the definition and, by doing so, help them understand the desired mathematics more deeply. Engaging with logic can aid mathematical learning; such explorations, which Abbott regularly incorporates, are ways this teaching practice is exemplified in learning real analysis.

References

1. Abbott, S. (2015). *Understanding analysis* (2nd ed.). New York, NY: Springer.
2. Buchberger, B. (1989). Should students learn integration rules? *SIGSAM Bulletin, 24*(1), 10–17.
3. Common Core State Standards in Mathematics (CCSSM). (2010). Retrieved from: http://www.corestandards.org/the-standards/mathematics

Algebraic Limit Theorems and Error Accumulation

4

4.1   Statement of the Teaching Problem

Students sometimes have to work with numbers whose decimal expansions do not terminate. To make such numbers manageable, they often round these values. (Although sometimes these are truncated and not rounded values, the issues are the same and we use the term rounded throughout the chapter.) For instance, the decimal expansion of $4\sqrt{5}$ is $8.9442719099\ldots$, but students often round to write this as 8.94. In general, students feel more comfortable working with values they regard to be "numbers"—like the counting numbers or short decimal representations, whereas they feel less comfortable with expressions like $4\sqrt{5}$ or long strings of decimals. This is understandable. Decimal notation is familiar and ubiquitous, and rounded values are easy to operate with and to record.

There are many instances when teachers actually want students to round, but at what point in a computation is rounding most appropriate? A general rule is that students should round at the end of a computation, but what goes wrong when students round in the middle? Often very little.

Consider the following pedagogical situation:

A student, Adrian, sets up and solves the equation,

$$\sin(59°) = \frac{x}{4\sqrt{5}}$$

by showing the following work:

$$0.85 = \frac{x}{8.94}, \text{ so } x = 0.85 \cdot 8.94 = 7.59.$$

(continued)

N. H. Wasserman et al., *Understanding Analysis and its Connections to Secondary Mathematics Teaching*, Springer Texts in Education, https://doi.org/10.1007/978-3-030-89198-5_4

> The teacher, Mr. Lee, walks around the room and observes the student's work. Mr. Lee tells Adrian:
>
> > Remember, do not round in the middle of the problem—wait until the end.
>
> Adrian objects to this remark:
>
> > Well, my answer is basically the same as Veronica's. She got 7.67, and she rounded only at the end. I finished faster and I understand my way better anyway.

Mr. Lee's advice is sound—it is generally better not to use approximated values in calculations if you are able to use more precise ones. In this sense, the teacher has responded fairly by pointing out that rounding should occur at the end. But the student offers two counterarguments: the difference compared to the actual answer is relatively small, and his solution method was faster to compute and easier for him to understand. These counterarguments are legitimate; there is a need to balance demands for accuracy with other practical concerns. While Adrian's approach holds up in this case, there are mathematical constraints around the utility of his approach (TP.1). So how does a teacher respond? Are there effective ways to illustrate that the student's approach might be problematic in the general case? What are some practical ways to respond to this sort of reasonable push-back from students?

Before moving on, think about how you, as a teacher, might respond to the student in this pedagogical situation.

4.2   Connecting to Secondary Mathematics

4.2.1   Problematizing Teaching and the Pedagogical Situation

In this section, we problematize three potential responses to the student regarding the issue of rounding.

One possible response would be to agree with the student—to regard the issue of when to round as not problematic. An argument for this response could be made in relation to the teacher's mathematical aims. Perhaps the goal of the problem is solving equations, for which the student's work demonstrates good algebraic reasoning. Multiplying both sides of the equation by the same value produces an approximate solution for x. The student's solution in this case demonstrates understanding of the intended mathematics. A drawback of this approach is that "attending to precision" is part of mathematical practice (e.g., [2]). Because a more precise answer *exists* it should probably be used. The geometric context of this particular example also comes into play: when solving for missing side lengths of a right triangle, students can check their results through other relationships such as the Pythagorean Theorem. Too much imprecision could lead to inaccurate conclusions. Furthermore, even in an algebraic context such rounding might be undesirable. Imagine a student who solves the equation $7/3 = \frac{x}{6/7}$ using the decimal approximations $2.33 = \frac{x}{0.85}$, and obtains an answer of 1.9805 rather than 2. Such instances highlight the advantages of encouraging students to work with less-preferred representations of numbers like fractions. They also draw attention to the need to understand how the impact of rounding early on in a computation affects the end result.

A second response is to declare the student's solution incorrect and mandate that students avoid rounding until the end of their computations. In some ways, this simplifies the situation. It provides clear expectations for students, which can be good in teaching. However, in this case, the rule is presented arbitrarily and without an accompanying reason. To practice TP.5 means to avoid giving rules without providing an explanation. Any consequences given for not adhering to this rule may also feel artificial given the close proximity of Adrian's answer to Veronica's.

A third possible response would be to superimpose a real-world context onto the issue of rounding. A teacher might claim that in designing a spaceship even very small errors in the real world can have tremendously negative consequences. The response here focuses on making the modest discrepancy in the answer "feel" more consequential. Yet, this line of reasoning still has challenges. This argument relies on convincing students that small errors can have large effects in applied settings, but students might be skeptical. They could insist that being off by a few tenths is not problematic in most situations; or they might say that while this would be true of an engineer designing spaceships, they are not engineers designing spaceships but students in a mathematics class. Perhaps the more important point is that such a response still does not illuminate the fact that rounding can result in very large errors—it only tries to make small errors feel large. In this sense, the response only partially addresses the student's counterargument.

4.2.2   Approximation and Error Accumulation

A rounded number is an approximation. This means we can think of error as we did in Chap. 3. In particular, if a_{appr} is a rounded approximation of a number a then we can consider both the *actual error*, e_{appr}, defined by

$$e_{appr} = \left| a_{appr} - a \right|$$

and the *potential error* or *error-bound*, e, which satisfies

$$\left| a_{appr} - a \right| < e.$$

Recall from Chap. 3 that the inequality $\left| a_{appr} - a \right| < e$, can be understood with two different referents. We might use a as the referent point, in which case the statement gives us the locus of points on the number line where a_{appr} is located; or we might use a_{appr} as the referent point, in which case it tells us the range of values where a is located. Both are depicted below.

$$a - e < a_{appr} < a + e \qquad\qquad a_{appr} - e < a < a_{appr} + e$$

Because the potential error e is a *bound* on the actual error e_{appr}, there are many possible values for e but only one for e_{appr}. It also means a value for e tends to be more readily accessible than one for e_{appr}. As an example, the potential error of a rounded decimal approximation can be computed from the number of decimal places—0.3 approximates the fraction $\frac{1}{3}$ to the tenths place, meaning we can use $e = 0.1$. (Note that this is indeed an upper bound for $e_{appr} = \left|0.3 - \frac{1}{3}\right| = \frac{1}{30}$.) Likewise, 3.14 approximates π to the hundredths place, meaning we can use $e = 0.01$, which produces the bound $3.13 < \pi < 3.15$. (Here, e_{appr} is the irrational number $\pi - 3.14$.)

In this chapter we consider not just individual approximations, as we did in Chap. 3, but what happens when we *operate* on approximated values. We consider how the error in the initial approximations accumulates, or changes, when the approximations are algebraically combined. To continue the above example, let $a = \pi$ and $b = \frac{1}{3}$. Using the notation e_a for the error-bound of a, we see that $a_{appr} = 3.14$ comes with error-bound $e_a = 0.01$. For $b = \frac{1}{3}$, the approximation $b_{appr} = 0.3$ has error-bound $e_b = 0.1$. What happens when we use these approximations to compute the sum $\pi + \frac{1}{3} \approx 3.14 + 0.3 = 3.43$? How far off could this approximated sum, 3.43, be from $\pi + \frac{1}{3}$? The potential error inequalities $|3.14 - \pi| < 0.01$ and $\left|0.3 - \frac{1}{3}\right| < 0.1$ can be arranged as

$$\pi - 0.01 < 3.14 < \pi + 0.01$$

$$\tfrac{1}{3} - 0.1 < 0.3 < \tfrac{1}{3} + 0.1$$

and summing yields

$$\left(\pi + \tfrac{1}{3}\right) - 0.11 < 3.43 < \left(\pi + \tfrac{1}{3}\right) + 0.11.$$

The approximate sum 3.43 must be within 0.11 of the actual sum. That is, the new potential error that results from adding two approximations is no worse than the sum of the initial potential errors, $0.01 + 0.1 = 0.11$. The addition of approximated values can be visualized as a linear transformation on a number line.

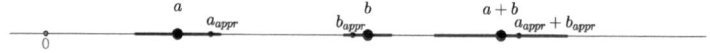

If e_a is the radius of the interval centered at a and e_b is the radius of the interval centered at b, then the interval centered at $a + b$ has radius $e_a + e_b$.

The same kinds of questions arise if we operate on approximations in other ways, such as subtracting, multiplying, or dividing them. To capture this idea of error accumulation more generally we give the following definition:

Definition For approximations a_{appr} of a and b_{appr} of b, each having potential errors e_a and e_b, **error accumulation** refers to the new potential error $e_{a \oplus b}$ that results from doing some operation ($\oplus$) to a_{appr} and b_{appr}.

Returning to the teaching scenario where the student solved for x by computing $x = 0.85 \cdot 8.94 = 7.59$, we can reframe the student's work as one of multiplying two approximated values. The potential error for both approximations (e_a and e_b) is 0.01. But what about the computed product, 7.59? What is *its* potential error ($e_{a \cdot b}$)? The actual answer, a little less than 7.67, suggests this new error has to be at least 0.07. Is there a way to calculate this error-bound from the potential errors for each factor? Is there a general method for calculating the error accumulation that results from other kinds of algebraic combinations?

It is to these issues that we turn next.

4.3   Connecting to Real Analysis

To connect this discussion to a real analysis course we return to the analogy introduced in Chap. 3 between the potential error inequality $\left| a_{appr} - a \right| < e_a$ and the expression $|a_n - a| < \varepsilon$ which appears in the definition for convergent sequences. For a sequence $(a_n) = (a_1, a_2, a_3, \ldots)$ that converges to a, we imagine the terms a_n getting closer to a as n gets large. For our purposes we want to think of each a_n in the sequence as an approximation of a (i.e., a_{appr_n}), so that the expression $|a_n - a| < \varepsilon$ can be interpreted to say that the approximation a_n has error-bound ε.

To study how error accumulates when we algebraically combine approximations, it turns out we can use the theorems from analysis that explain what happens when we algebraically combine convergent sequences.

4.3.1   The Algebraic Limit Theorem for sequences

The Algebraic Limit Theorem for sequences asserts what happens to convergent sequences when we add, multiply, or divide them (cf., Theorem 2.3.3 in Abbott [1]):

Theorem (Algebraic Limit Theorem) Let $\lim a_n = a$ and $\lim b_n = b$. Then,

1. $\lim (c \cdot a_n) = c \cdot a$, for all $c \in \mathbb{R}$
2. $\lim (a_n + b_n) = a + b$
3. $\lim (a_n \cdot b_n) = a \cdot b$
4. $\lim (a_n/b_n) = a/b$, provided $b \neq 0$.

The Algebraic Limit Theorem confirms that, when algebraically combining convergent sequences, things go as we might expect. For instance, if (a_n) converges to a and (b_n) converges to b, then the new multiplied sequence $(a_n b_n)$ converges to ab. The agenda of the real analysis proof is showing that the potential error of the new combined sequence can be made arbitrarily small. We do not provide the proofs (they can be found in Abbott's Theorem 2.3.3.), but we do list the four key inequalities that form the cornerstone for the proofs of each part of the Algebraic Limit Theorem. The numbering below corresponds to the numbering in the statement of the theorem:

1. $|ca_n - ca| \leq |c| \cdot |a_n - a|$
2. $|(a_n + b_n) - (a + b)| \leq |a_n - a| + |b_n - b|$
3. $|(a_n \cdot b_n) - (a \cdot b)| \leq |b_n| \, |a_n - a| + |a| \, |b_n - b|$
4. For N_1 sufficiently large such that, for all $n \geq N_1$, b_n is closer to b than to 0,
   then: $\left| \frac{1}{b_n} - \frac{1}{b} \right| \leq \frac{2}{|b|^2} \cdot |b_n - b|$.

(Note inequality (4) is really about the reciprocal $1/b_n$ rather than the quotient a_n/b_n.) These inequalities are central to proving the algebraically combined sequences converge to their respective limits, but they can also be adapted to our particular agenda of estimating the error accumulation of combined approximations. Each inequality is true for every term in the corresponding sequence, so if we suppose a_n and b_n are our approximations a_{appr} and b_{appr}, then these statements tell us something about error accumulation when a_{appr} and b_{appr} are combined in each way.

4.3.2   Implications for Error Accumulation

Take a look at each of the four inequality statements.

The left-hand side of the inequalities all have a similar form—the difference between an operated-on approximation and its theoretical value (e.g., $|ca_n - ca|$). In fact, they are all statements about potential error in our operated-on approximations. In particular, they indicate the accumulated error in the operated-on approximation is no worse than the expression on the right-hand side. The expressions on the right-hand are all in terms of the initial error of each approximation, $|a_n - a|$ and $|b_n - b|$. That is, we can interpret each inequality as a statement about how initial errors accumulate when operating on approximations.

Claim For approximations a_{appr} of a and b_{appr} of b, each having potential errors e_a and e_b:

1. the error accumulation of the **scalar product** ca_{appr} is no worse than initial error scaled by $|c|$; i.e., $e_{ca} = |c| e_a$,

2. the error accumulation of the **sum** $a_{appr} + b_{appr}$ is no worse than the sum of the initial errors; i.e., $e_{a+b} = e_a + e_b$,
3. the error accumulation of the **product** $a_{appr} \cdot b_{appr}$ is no worse than the sum of the initial error of a scaled by $|b_{appr}|$ and the initial error of b scaled by $|a|$; i.e., $e_{ab} = |b_{appr}|e_a + |a|e_b$,
4. the error accumulation of the **reciprocal** $\frac{1}{b_{appr}}$ is no worse than the initial error scaled by $\frac{2}{|b|^2}$; i.e., $e_{1/b} = \frac{2}{|b|^2}e_b$.

Re-read each inequality statement and the corresponding inequality statement from Sect. 4.3.1 and convince yourself that they mean the same thing. We will refer to these claims as "rules" for error accumulation. Notably, the sum rule arrived at in statement (2) aligns with the conclusions we found previously. Statement (4) is about the reciprocal; but by writing $\frac{a_{appr}}{b_{appr}}$ as $a_{appr} \cdot \frac{1}{b_{appr}}$, Problem 4.6 asks you to derive the corresponding quotient rule. And Problem 4.7 asks you to consider how some of these rules about error accumulation might be simplified under further assumptions.

To get a better sense of these rules let's return to our previous example.

Example Suppose we approximate π with 3.14 and $\frac{1}{3}$ with 0.3. How much potential error is there in: (i) $5 \cdot 3.14$; (ii) $3.14 \cdot 0.3$; (iii) $\frac{1}{3.14}$; (iv) $\frac{0.3}{3.14}$?

The rules provide a bound for how error *potentially* accumulates when operating on approximated values. For (i), $5 \cdot 3.14 = 15.7$ is an approximation for 5π. The scalar product rule in (1) states that the potential error is no worse than $|5| \cdot 0.01 = 0.05$—i.e., that 15.7 is within ± 0.05 of 5π. This is sensible. The rule says error potentially accumulates up to five times the original error, or $e_{5a} = 5 \cdot e_a$.

For (ii), $3.14 \cdot 0.3 = 0.942$ is an approximation for $\frac{1}{3}\pi$. From the product rule in (3), the potential error is no worse than

$$|3.14| \cdot 0.1 + |1/3| \cdot 0.01 \approx 0.3173.$$

For comparison, the actual error in this case is about 0.1052. Note the use of one approximated value, 3.14, and one theoretical value, 1/3, in the error accumulation calculation above. If a theoretical value for, say, a is not available we can overestimate it with $|a_{appr}| + e_a$.[1] In most cases, the errors of the initial approximations (e_a and e_b) are relatively small compared to the approximations (a_{appr} and b_{appr}) and so swapping theoretical values for approximated ones in the rules causes very little change to our error estimates. This means that when multiplying two approximated numbers, as a rule of thumb the error of the product accumulates by approximately $|a|$ times the error in b plus $|b|$ times the error in a.

---

[1] In general, we can replace statement (3) in the claim with the more conservative estimate, $e_{ab} = |b_{appr}|e_a + (|a_{appr}| + e_a)e_b$.

For (iii), the reciprocal of an approximation, $\frac{1}{3.14} \approx 0.31847$, has a potential error no worse than $\frac{2}{|\pi|^2} \cdot 0.01 \approx 0.002$. Here, the error has gotten smaller because it was scaled by $\frac{2}{|\pi|^2}$, which is a value between 0 and 1.[2]

Lastly, for (iv), $0.3/3.14 = 0.3 \cdot \frac{1}{3.14} \approx 0.0955$ approximates $\frac{1}{3\pi}$. Combining the product and the reciprocal rules, we can estimate the error in the quotient as

$$\left(\left|\frac{1}{3.14}\right| \cdot 0.1\right) + \left(\left|\frac{1}{3}\right| \cdot \frac{2}{|\pi|^2} \cdot 0.01\right) \approx 0.0325.$$

4.3.3 Visualizing the Potential Error Inequality for Products

In a proper proof of the Algebraic Limit Theorem, the primary conclusion is that although error accumulates, it does not do so uncontrollably. Even though operating on sequences might increase the error, the potential error at each stage is bounded by some knowable combination of the original errors. This means we can go out far enough in the new operated-on sequence so that the accumulated error is arbitrarily small (since we know the original errors become arbitrarily small). The theorem tells us that accumulated error converges to zero as n increases in the sequence. In the context of our discussions about approximations, we have borrowed the parts of the proof that tell us *how* the errors interact when approximations are operated upon and fashioned them into rules for error accumulation.

The derivation of the four inequalities that underlie our accumulation rules all make use of the triangle inequality. They are not especially difficult (see Abbott, Sect. 2.3), but the product statement (3) appears a bit mysterious. The product accumulation rule $e_{ab} = |b_{appr}|e_a + |a|e_b$ is based on the inequality

$$|a_{appr}b_{appr} - ab| \leq |b_{appr}||a_{appr} - a| + |a||b_{appr} - b|.$$

To visualize this inequality consider the area model in Fig. 4.1. The products $a_{appr}b_{appr}$ and ab appear as the areas of two large shaded rectangles. The absolute value of their difference $|a_{appr}b_{appr} - ab|$ is, at most, the area of what's left of these two rectangles when we remove their intersection. In the figure, this remaining area appears as the tall thin rectangle on the right with dimensions $b_{appr} \times (a_{appr} - a)$ and the long thin rectangle across the top with dimensions $a \times (b_{appr} - b)$. The fact that the sum of these two areas is an upper bound for $|a_{appr}b_{appr} - ab|$ verifies the original inequality statement.

---

[2] If we do not have an exact value for $|b|$, we can use the more conservative estimate, $e_{1/b} = \frac{2}{(|b_{appr}| - e_b)^2} e_b$.

Fig. 4.1 Area model for the products ab and $a_{appr}b_{appr}$

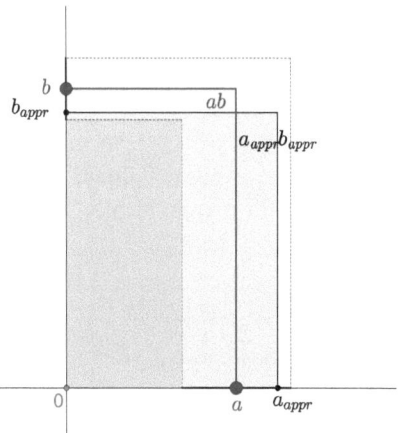

4.4 Connecting to Secondary Teaching

In the initial teaching situation, the student's use of approximated values is not intrinsically problematic, and the teacher's advice of not rounding until the end is also sound. The student's reasonable counterarguments in the scenario emphasize that teachers should be prepared to give justifications for their advice. Simply stating a rule—"Remember, do not round in the middle of the problem, wait until the end"—without any sort of mathematical justification runs contrary to TP.5. A response to the student in this teaching situation can be informed by the insights from the real analysis proofs of the Algebraic Limit Theorem.

4.4.1 Applying Principles of Error Accumulation to Design Problems

In the original problem, the student substituted rounded values for $\sin(59°)$ and $4\sqrt{5}$ with a potential error of 0.01 for both approximations. Using these two values, the student solved the equation by multiplying them. The student's answer of 7.59 was reasonably close to the actual answer of about 7.67. The product rule from this chapter can be used to calculate the potential error accumulated in the student's calculation. Specifically, the error in the student's approximated answer is no worse than $0.85 \cdot 0.01 + 4\sqrt{5} \cdot 0.01 \approx 0.098$, which is about 10 times the initial error.

Although it is useful to determine the potential error in the student's answer, we regard it as more important to think about how the teacher might *apply* the ideas about error accumulation to respond to the student. Rather than simply telling the student the error could get big, it would be more beneficial to construct another problem for the student—one that demonstrates that rounding early in a computation can result in a relatively large error. Proceeding in this way is aligned with TP.2, using special cases to illustrate mathematical ideas.

Consider the following continuation of the teaching situation:

> Mr. Lee responds to Adrian: "I would like you to try your rounding approach on the following problem:
>
> $$\sin(59°) = \frac{x}{360\sqrt{5}}$$
>
> Tell me, how close is your answer this time?"

The changes that have been made to the problem appear to be minor, but they make an important difference. The student's approach presumed the potential error in both approximations to be the same; i.e., $e_a = e_b = e$. With this assumption, the error accumulation for a product simplifies: the original error e will accumulate by no more than a factor of $(|a| + |b_{appr}|)$. That is to say, the initial potential error of 0.01 will grow approximately by a factor of the sum of the two values being multiplied. The seemingly minor change of replacing $4\sqrt{5}$ with $360\sqrt{5}$ in the calculation is pedagogically-motivated—it is meant to increase the accumulated error. By changing the values in the equation, the potential error in solving for x in this new equation grows by a factor of $(0.85 + 360\sqrt{5}) \approx 805.83$, which is 800 times the original error! An initial rounding to the hundredths place could result in an error of more than 8. (Alternately, Mr. Lee could edit the equation to use $\tan(59°)$ since, unlike sine, the tangent function is not bounded.) In fact, Adrian's error in this new problem would be about 5.77—a difference most students would regard as non-trivial.

This example can be adapted to a wide class of problems. Students invoke decimal approximations when solving other equations with rational coefficients, such as $\frac{2}{7}x = 60\frac{6}{7}$. Solving for x in this example involves the quotient of two approximations. As is evident from the reciprocal and product accumulation rules, larger values for a or smaller values for b result in increased potential errors. This particular example has both. So the actual error ends up being 400 times the original error (presuming a student has rounded both numbers to the same number of decimal places)—a large error indeed!

In these examples, we have applied the rules for error accumulation to construct new problems, intentionally adjusting the values being approximated so as to increase the potential error accumulation in the solution. This is an example of "using a special case to illustrate a mathematical idea" (TP.2). We have designed an exercise to illustrate that using rounded values instead of actual values can lead to large errors. Special cases are important in mathematics, and they also serve a pedagogical purpose. In this case we wanted to convince a student their rounding approach could be problematic. Our example does not communicate to the student precisely when rounding leads to large errors—that's a heavier lift that requires engaging the ideas in the real analysis proofs—but it does supply a cautionary warning that rounding can be problematic.

The student's approach to rounding has some limitations, and the various constructed exercises discussed are designed specifically to reinforce the teacher's maxim to not round until the end of the problem. TP.5 suggests that teachers avoid giving rules without an accompanying mathematical explanation. In this scenario, the explanation takes the form of an exercise rather than a verbal justification. Observing the large errors that accumulate in the constructed exercises is a compelling piece of evidence in favor of the teacher's advice and may in fact be more convincing than any words the teacher could say. Providing students opportunities to *experience* ideas and not just have them *explained* is an important part of teaching.

As a final comment, we address the question of how the more advanced content of real analysis relates to the daily reality of teaching secondary school mathematics. In this chapter we have seen how the proofs for the various parts of the Algebraic Limit Theorem contain insights for understanding the nature of error accumulation. Knowledge of the ideas from the proofs empowers a teacher to engage student questions and counterarguments about the efficacy of approximations with carefully crafted examples designed to illuminate certain pitfalls. The constructed examples showcase a teacher applying ideas learned in real analysis to respond to a student but in a way that does not involve an exposition of more advanced concepts. This suggests that real analysis can be an impactful subject for teachers in ways that do not amount to teaching gifted secondary students proofs for results like the Algebraic Limit Theorem. Despite its formal reputation, analysis represents a body of ideas that can reveal new insights about day to day issues that arise in teaching.

4.4.2 The Tip of the Iceberg

Throughout this chapter we have focused on potential error rather than actual error. This is the more useful and appropriate point of focus. The fact that we are approximating suggests that there is some uncertainty in the theoretical value being approximated. This means the actual error is not typically known—or even knowable. Our rules for error accumulation are based on overestimates, or worst-case scenarios, so it is certainly possible that actual errors might decrease even as our potential error calculations increase. For example, rounding $\frac{1}{3} + \frac{2}{3}$ gives $0.33 + 0.67$, where each approximation has an error but the sum is perfect. In this case, it helps to give the actual error a *signed direction*. When we add, the actual errors cancel out; the potential error estimates of course do not. The triangle inequality—which is at the root of all our error accumulation rules—is an equality if and only if the errors have the same sign. This explains in part why the actual accumulated error is likely to be smaller than the potential accumulated error. The mix of positive and negative terms results in some cancellation that yields a better than expected approximation.

The calculation of the potential error meanwhile assumes the worst case where the errors stack up in one direction.

Another factor contributing to an inflated accumulated error estimate is an overestimate in the original error. Taking the rounded value of 3.14 as our estimate of π, we have been using $e = 0.01$ as our error-bound but the actual error is closer to 0.00159. Generally speaking, if our initial error-bounds are close to the actual errors, and all the actual errors have the same sign, then the error accumulation formulas tend to give values close to the actual error of the final computation.

Another direction for further investigation is how errors behave in cases beyond the scope of the Algebraic Limit Theorem. The error accumulation rules developed in this chapter apply to the scalar product, sum, product, and quotient. From these we could derive a rule for what happens to our error when we square a_{appr}^2 or cube a_{appr}^3 an approximation. New tools are required, however, to sort our how error accumulates when we take the square root $\sqrt{a_{appr}}$ or apply a function like $\sin(a_{appr})$ or $\tan(a_{appr})$. These sorts of questions are studied in depth in courses on numerical analysis, but preliminary error estimates can be derived from ideas in a real analysis course (and the theorems about continuity in particular).

This chapter is just the tip of the iceberg in terms of understanding error accumulation from approximations.

Problems

4.1 A student approximates 12/7 as 1.714, and 7/6 as 1.167. Use the initial potential errors (0.001 for each), and the rules about error accumulation in this chapter, to determine a bound for the potential error if the student used those approximations to compute: (i) $\frac{12}{7} + \frac{7}{6}$; (ii) $\frac{12}{7} \cdot \frac{7}{6}$; and (iii) $\frac{12}{7} \div \frac{7}{6}$. What is the actual error in each case?

4.2 In an algebra class, students are solving for the roots of the quadratic, $f(x) = x^2 - 2x - 27$. Students use the quadratic formula to find the roots to be at $x = 1 \pm 2\sqrt{7}$, and then evaluate the quadratic formula using their calculator. The calculator uses an approximation for $\sqrt{7}$ that is accurate to eight decimal places—i.e., the potential error in the calculator's approximation is 0.00000001. How much error could be introduced in the calculator's evaluation of the roots of the quadratic?

4.3 During class, a teacher recommends that students approximate π with the value 3.14 for computations. Describe a specific situation in secondary mathematics where such an approximation might lead to a large error.

4.4 (i) Design a problem of the form $Ax + B = C$, with $A, B, C \in \mathbb{R}$, to be given to the students as a multiple choice item, for which a student using the "round to the nearest hundredth" approach would almost certainly select the incorrect choice. (Your problem should include the multiple choice options.) (ii) Assuming the potential error is $e = 0.01$ for any decimal approximation, provide an analysis of the *potential error* introduced in solving the equation with rounded values. (iii) Provide an analysis of the *actual error* for the solution with rounded values in comparison to the theoretical solution, $x = \frac{C-B}{A}$. Discuss your multiple choice options in relation to this error.

4.5 A teacher gives a question in class that involves determining the perimeter and the area of a rectangle where the side lengths are $\sqrt{75}$ and $\sqrt{362}$. The teacher writes on her answer key (rounding at the end), $P = 55.37$ and $A = 164.77$. A student's calculator shows $\sqrt{75} = 8.660254038$, and $\sqrt{362} = 19.02629759$. If the student approximates these two numbers before making the perimeter and area computations—presume the student is simply "truncating" throughout the problem—use your knowledge of how error accumulates to determine the degree of accuracy that would be required for the student to get the *same* answer as the teacher. That is, should the student's original rounding for the square roots be accurate to the tenths, hundredths, thousandths, etc.? Explain whether the requisite accuracy level differs between the perimeter and area problems, and why.

4.6 Use the fact that $\frac{a_{appr}}{b_{appr}} = a_{appr} \cdot \frac{1}{b_{appr}}$ to determine a general rule for how error accumulates for a quotient of two approximations. You will need to combine how error accumulates for both reciprocals and products.

4.7 Suppose we make some additional assumptions about our approximations: (i) the theoretical values (and their approximations) are *positive* ($a, b, a_{appr}, b_{appr} > 0$); (ii) the initial potential errors are the *same* ($e_a = e_b = e$); and (iii) $a_{appr} < a$ and $b_{appr} < b$ (our approximations are *under-approximations*—such as truncating a decimal expansion). These three assumptions simplify some of the rules about error accumulation. (i) For products, we had $e_{ab} = |b_{appr}|e_a + |a|e_b$. With these additional assumptions, what is the new claim about the error accumulation of a product? (ii) In the reciprocal inequality, we have: $\left|\frac{1}{b_{appr}} - \frac{1}{b}\right| = \frac{1}{|b||b_{appr}|} \cdot |b_{appr} - b|$. With these additional assumptions, what is the new claim about reciprocals? (Note: you should no longer have a '2' in the numerator.) (iii) Building on (ii), with these additional assumptions, what is the new claim about the error accumulation of a quotient?

4.8 Suppose we allow our actual errors to be signed (positive or negative). That is, we define $a_{appr} = a + e_{appr_a}$ and $b_{appr} = b + e_{appr_b}$. Use substitution to show the product $a_{appr} \cdot b_{appr}$ has an error of $\left(a e_{appr_b} + b_{appr} e_{appr_a}\right)$ from ab, and relate this to the product rule in Sect. 4.3.2.

4.9 In Exercise 2.3.7 in Abbott's text, he asks for students to "give an example" (or argue that such a request is impossible) of, for example, "sequences (x_n) and (y_n), which both diverge, but whose sum $(x_n + y_n)$ converges." Indeed, in many sections, Abbott uses exercises similar to this. Describe what teaching principle you believe is exemplified in these exercises, and explain your reasoning.

4.10 Chapter 2 in Abbott's text is broadly about defining and understanding *sequences*. But at the beginning of the chapter (Sect. 2.1), and then later at the end of the chapter (Sect. 2.7), Abbott explicitly talks about infinite *series*. In the introduction to the chapter, Abbott uses an example: $1 - \frac{1}{2} + \frac{1}{3} - \frac{1}{4} \ldots$. Then, he states, "The crucial question is whether or not properties of addition and equality that are well understood for finite sums remain valid when applied to infinite objects such as [the example]" (p. 40). Describe what teaching principle you believe is exemplified in his text.

References

1. Abbott, S. (2015). *Understanding analysis* (2nd ed.). New York, NY: Springer.
2. Common Core State Standards in Mathematics (CCSSM). (2010). Retrieved from: http://www.corestandards.org/the-standards/mathematics

Divergence Criteria and Logic in Communication

5.1   Statement of the Teaching Problem

Euclid's *Elements* [2] is the most influential mathematics textbook in history. With a short set of postulates and some common notions and definitions, Euclid deduced thirteen volumes of geometrical propositions. His approach was axiomatic; it applied *logical* principles, which are now central to mathematics. The challenge is that teaching mathematics does not, and cannot, operate purely within this logical structure.

Teaching is an act of bridging between what a learner knows and does not yet know. To achieve this, teachers must communicate ideas in everyday language that students understand rather than in a strictly formal mathematical language. Finding a balance between these two modes of communication can be challenging. Consider the following two statements: "A square is a regular quadrilateral" and "A square is a rectangle." Grammatically, these follow an identical structure ("a [blank] is a [blank]"). Before reading on, think about how you interpreted them. What do you think the first one means? What about the second?

Logically, the two statements are different. In the first, "is" represents a *bidirectional* ($\Longleftrightarrow$) relation, which characterizes a definition or an 'if-and-only-if' statement. In the second, "is" represents a *directional* ($\Longrightarrow$) relation, which is used to articulate a property or an 'if-then' statement. A description of squares is that they are all rectangles, but the relationship is *not* true in reverse. For a different example, consider the statement: "You can have dessert if you finish your dinner." This is intended to communicate that the only way to have dessert is to finish your dinner. But this is an 'if-and-only-if', or bidirectional, interpretation (Dessert $\Longleftrightarrow$ Dinner). Technically, the grammatical phrasing uses an 'if-then' structure (Dinner $\Longrightarrow$ Dessert). In this literal interpretation we cannot be sure what happens if you do not finish your dinner. You may or may not get dessert—all we know is what happens if you *do* finish your dinner. Indeed, not attending to the

N. H. Wasserman et al., *Understanding Analysis and its Connections to Secondary Mathematics Teaching*, Springer Texts in Education, https://doi.org/10.1007/978-3-030-89198-5_5

direction of directional statements is commonplace in normal conversation; some might actually interpret the statement to mean the converse (Dinner $\Longleftarrow$ Dessert). This is because logical relationships can get lost in the translation to everyday language—a translation which is a necessary part of teaching.

Consider the following pedagogical situation:

> A geometry teacher, Ms. Rojas, has been teaching students about special quadrilaterals. One of those special quadrilaterals, a trapezoid, was defined to be "a quadrilateral with exactly one pair of parallel sides."
>
> Ms. Rojas is now discussing isosceles trapezoids. Throughout her explanations and in her responses to student questions, Ms. Rojas makes the following three statements:
>
> 1. "A trapezoid is isosceles if the non-parallel opposing sides are congruent"
> 2. "An isosceles trapezoid is a quadrilateral with congruent diagonals"
> 3. "A trapezoid is isosceles if the diagonals are congruent"

In these three statements, Ms. Rojas appears to give a definition for an isosceles trapezoid as well as some properties. The language she uses is not overly formal, and the fact that she describes the main ideas in multiple ways is a positive aspect (TP.6). But one of the challenges of teaching is being aware of the intended mathematical relationships, the way those relationships are expressed, and the possible ways they might be interpreted by students in the class. This is especially important when discussing relationships that are not logically equivalent.

Before moving on, think more about each statement. How could you rewrite each one using a more formal logical structure, and would the meaning that arises from that structure align with what you believe the teacher is trying to communicate?

5.2   Connecting to Secondary Mathematics

5.2.1   Problematizing Teaching and the Pedagogical Situation

We problematize the three isosceles trapezoid statements by considering the different ways they may be interpreted, including some which convey potentially contradictory meanings.

The primary logical relationships we discuss in this chapter are conditional and biconditional statements:

Definition A **conditional** statement is *directional*, of the form 'if A then B', or $A \Longrightarrow B$, where A is referred to as the *condition* and B the *consequence*.

Definition A **biconditional** statement is *bidirectional*, of the form '*A* if-and-only-if *B*', or $A \iff B$, where *A* and *B* are interpreted as *equivalent*; that is, $A \implies B$ and $A \impliedby B$.

As discussed, a statement has possible logical meanings across each of three categories: the *intended* relationship, the *expressed* relationship, and the *interpreted* relationship. To frame the scope of the challenge, we could categorize different logical possibilities in the table below. We introduce the table not for the purpose of discussing every possibility, but for situating the few examples we do discuss.

Speaker		Listener		
Intended	Expressed	Interpreted		
		$\implies$	$\iff$	$\impliedby$
$\implies$	$\implies$			
$\implies$	$\iff$			
$\implies$	$\impliedby$			
$\iff$	$\implies$			
$\iff$	$\iff$			
$\iff$	$\impliedby$			

For example, the first row of the table represents when the speaker intends a directional statement and expresses a directional statement. The three blank cells to the right allow for the possibility that the listener interprets the statement accurately (first column), as a biconditional (second column), or in the reverse (third column).

In the first statement from the teaching situation, the teacher intended to give a definition. Definitions are necessarily biconditional statements—if we call something *A* then it has property *B*, and if we see something with property *B* we call it *A*. In this case, the definition is:

$$\text{Isos. Trap.} \iff \text{Non-parallel Opp. sides Cong.}$$

The teacher's statement, however, is expressed in the form of a directional relationship:

$$\text{Non-parallel Opp. sides Cong.} \implies \text{Isos. Trap.}$$

The intended relationship is $\iff$, but it has been expressed as $\implies$, which corresponds to row four in the table above.[1] A listener might interpret the relationship as it was expressed ($\implies$) which has a different logical meaning than was intended. The condition expressed is 'if Non-parallel Opp. sides Cong.' This is a *criterion* for isosceles trapezoids: any trapezoid that meets this condition is isosceles. However,

---

[1] This is very common in giving definitions in mathematics; indeed, some definitions given in this book also have been expressed in this way.

with the directional interpretation, we cannot conclude anything about trapezoids whose non-parallel opposing sides are *not* congruent—who knows, some of these might also be isosceles trapezoids. Directional implications tell us nothing about when the condition is not met. With this interpretation, we could not determine whether either of the following quadrilaterals is an isosceles trapezoid.

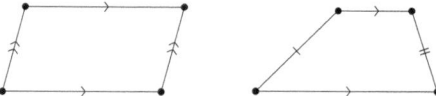

For the quadrilateral on the left, non-parallel opposing sides do not exist. For the quadrilateral on the right, non-parallel opposing sides do exist but they are not congruent. In our scenario, the teacher's definition of trapezoid rules out the parallelogram on the left, but statement (1), as expressed, leaves ambiguous whether the trapezoid on the right is isosceles. We refer to this as the **'iff' confusion**: a biconditional relationship is expressed as a conditional one, and so it is difficult to interpret whether the intended relationship goes both ways.

Now consider statement (2), which seems to describe a property of isosceles trapezoids: their diagonals are congruent. But the statement uses the word "is". Interpreted as a conditional statement, we get:

$$\text{Isos. Trap.} \implies \text{Quad. with Cong. Diag.}$$

This interpretation matches the actual relationship. The problem is that the word "is" can also be interpreted to indicate a biconditional relationship:

$$\text{Isos. Trap.} \iff \text{Quad. with Cong. Diag.}$$

Indeed, the grammatical structure of the statement—an isosceles trapezoid is not just a quadrilateral but one with congruent diagonals—matches how definitions are often given (e.g., a square is not just a quadrilateral, but a particular kind, a regular one). Interpreted this way, the statement is false. Two counterexamples, quadrilaterals that have congruent diagonals but are not isosceles trapezoids, are shown below.[2]

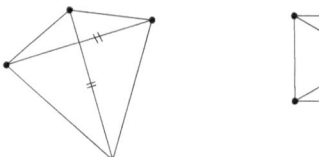

---

[2] The rectangle is an unusual case, explored more in the Chap. 6. For now it suffices to say that, according to the definition given in class, the rectangle would not be a trapezoid at all.

The logical meaning of statement (2) hinges on the interpretation of "is," which might express an intended relationship of $\Longrightarrow$, or $\Longleftrightarrow$. We refer to this as the **'is' confusion**: a logical connector expressed by "is" makes interpretation difficult because the actual relationship could be conditional or biconditional.

Statement (3) can be paired with statement (1) because, grammatically, they have the same structure (which raises the same 'iff' concerns). It can also be paired with statement (2) because the teacher is communicating about congruent diagonals in both statements. Upon closer inspection, however, statement (3) is distinct because the property of congruent diagonals is part of the condition, which is the *reverse* of the second statement. To be honest, it is hard to say precisely what the intended condition of statement (3) actually is: it could be 'Quad. with Cong. Diag.' or possibly 'Trap. with Cong. Diag.'? Let's consider both:

$$\text{Isos. Trap.} \Longleftarrow \text{Quad. with Cong. Diag.}$$

$$\text{Isos. Trap.} \Longleftarrow \text{Trap. with Cong. Diag.}$$

With the first interpretation, the statement is false by the previous counterexamples. This interpretation would be the *converse* of statement (2) and, as we can see, expresses something logically distinct. The second interpretation is true. It indicates that congruent diagonals in a trapezoid is sufficient to conclude the trapezoid is isosceles. In conjunction with the second statement, it could serve as another way to define an isosceles trapezoid—a biconditional relationship, 'Isos. Trap. $\Longleftrightarrow$ Trap. with Cong. Diag.' If the teacher meant for statement (3) to have this latter interpretation, then its expressed directional structure does not capture the intended bidirectional meaning. If the teacher intended statement (3) to be a recasting of statement (2), then the teacher actually switched the condition; the intended relationship ($\Longrightarrow$) was expressed in the reverse ($\Longleftarrow$). This confusion could also happen during interpretation: something expressed as $\Longrightarrow$ being interpreted as $\Longleftarrow$. We refer to this as the **'converse' confusion**: a conditional (directional) relationship is expressed or interpreted in the reverse, switching the condition and the consequence.

As teachers who necessarily employ informal language to communicate formal mathematical ideas, it is important to be aware of how easily the intended logical relationships can get lost in the way they are expressed and interpreted.

5.2.2 Common Variants of Conditional Statements

There are several common variants of a conditional statement $A \Longrightarrow B$, some of which incorporate negation ($\neg$). Each are distinct, although pairs of them have the same truth value, meaning they are logically equivalent.

1. *Statement*: $A \Longrightarrow B$
2. *Converse*: $B \Longrightarrow A$

3. *Inverse*: $\neg A \implies \neg B$
4. *Contrapositive*: $\neg B \implies \neg A$

The converse was mentioned previously—reversing the condition and the consequence. This can confuse logical communication because, as we have seen, a statement and its converse are not logically equivalent. On the other hand, even though they look different, a statement and its contrapositive *are* logically equivalent. If A implies B, then not having property B implies not having property A.

Claim A conditional statement and its contrapositive are logically equivalent.[3]

5.3   Connecting to Real Analysis

The isosceles trapezoid statements highlight some of the challenges of mathematical communication, and these same kinds of challenges certainly arise in real analysis. In this section, we consider several statements about convergence of sequences. To connect this to the previous discussion, the strategy is to transport the same scrutiny about the expressed and intended mathematical relationships to a new domain. Whether the topic is geometry or advanced calculus, paying attention to logic can improve our understanding of the relationships—especially in the way directional statements communicate *descriptions of* or *criteria for* mathematical concepts.

5.3.1 Convergence Theorems

As an initial exercise, read through the following real analysis statements. Some are stated in an atypical way. As you do so, try to clarify the logical structure of the intended mathematical relationship ($\implies$ or $\iff$) and think about what other interpretations might be possible.

Real Analysis Statements

1. A sequence is convergent if all of its subsequences converge to the same limit
2. A convergent sequence is a sequence that's bounded
3. A sequence is convergent if it is monotone and bounded

---

[3] The converse and inverse have this same relationship, so they, too, are logically equivalent.

To showcase possible points of confusion, the three statements have been grammatically constructed to mirror the isosceles trapezoid statements. The content has changed, so the truth values might change as well, but notice that we are faced with the same ambiguity about the intended directional nature of each claim.

Statement (1) is true as a biconditional relationship. In this sense, the *'iff'* confusion applies since statement (1) is currently expressed as a directional statement. As written, the statement has the condition 'if All Subseq. Converg. to Same Lim.' and the consequence 'Seq. is Converg.' This is a true statement, albeit a rather trivial one since one of the subsequences would be the sequence itself. This may be the intended meaning but perhaps not. Interpreted this way, statement (1) says nothing about a sequence where the condition is not met. For instance, the sequence $(a_n) = \left(0, \frac{1}{2}, 0, \frac{3}{4}, 0, \frac{5}{6}, 0, \dots\right)$, depicted below, does not meet the condition since not all its subsequences converge to the same limit.

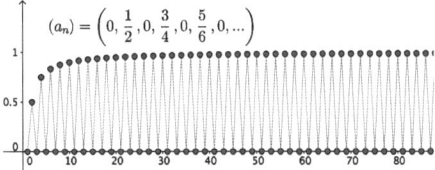

Does (a_n) converge? It does not, but statement (1) interpreted as a directional if-then proposition cannot be invoked to reach this conclusion. In fact, the relationship in statement (1) is biconditional, and this may be how it was intended. The converse proposition, 'if a sequence converges then all of its subsequences converge to the same limit' is a significant result in analysis that we discuss momentarily.

Statement (2) demonstrates the *'is'* confusion. The meaning changes depending on whether we interpret "is" as conditional ($\Longrightarrow$) or biconditional ($\Longleftrightarrow$). Interpreting statement (2) as a bidirectional statement means convergence and boundedness are equivalent—that 'Converg. $\Longleftrightarrow$ Bound.' This is not true. The sequence $(a_n) = \sin n$ is bounded between -1 and 1 but does not converge (see below). So it is not true that 'Converg. $\Longleftarrow$ Bound.'

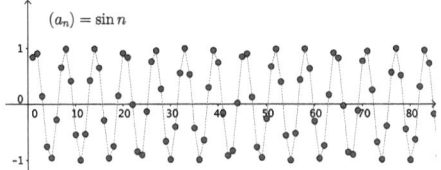

However, if we interpret "is" to be directional we get: 'Converg. $\implies$ Bound.' This is true. Convergence is the condition and boundedness is a *description of* convergent sequences—they all have this property.

Statement (3) grammatically mirrors statement (1), but like the first statement, the intended relationship is directional, not bidirectional. It also has similarities to statement (2) in that they both describe relationships between convergent sequences and bounded ones. Being a directional statement, the *'converse' confusion* might surface; one might mistakenly interpret the statement in the reverse: 'Converg. $\implies$ Mon. and Bound.' This implication is not true. Although we know convergent sequences must be bounded, they do not have to be monotone to be convergent. The sequence (a_n) where $a_n = \frac{\sin n}{n}$ converges to 0 (and hence is bounded), but it is not monotonic.

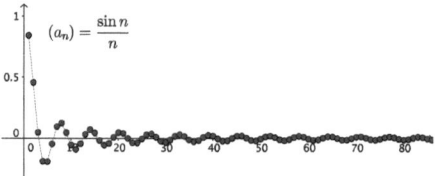

The intended statement is 'Converg. $\impliedby$ Mon. and Bound.' In contrast to statement (2), convergence is the consequence. Being monotonic and bounded is a sufficient *criterion for* convergent sequences.

These three statements can be recast as theorems that should be familiar from analysis (cf., Abbott's Theorems 2.3.2, 2.4.2, and 2.5.2). In this more formal setting, note how the statements are carefully crafted to clarify the logical relationships:

Theorems

1. **(Subsequence Convergence Theorem)** If a sequence converges, then all subsequences converge to the same limit.
2. **(Boundedness Theorem)** If a sequence converges, then it is bounded.
3. **(Monotone Convergence Theorem)** If a sequence is monotone and bounded, then it converges.

We do not provide proofs of these theorems (see Abbott's [1] text). Instead, we want to study their logical structure by exploring the inverse, converse, and contrapositives. The result is a robust investigation of the complementary notions of convergence and divergence, as well as a heightened clarity for the inner workings of propositional logic in general.

Before reading on, formulate the converse, inverse, and contrapositive of each theorem. In each case ask yourself, "Is this true?" and "What does this tell me about divergent sequences?"

5.3.2 Logical Implications About Divergence

Because a contrapositive statement is logically equivalent to the original, it often provides additional insight. Here, it supplies ways of understanding *divergent* sequences. The contrapositive of each theorem is given below. Because the original theorems are true, these statements are true as well.

Contrapositive Statements

1. **(Divergence Subsequences Criterion)** If not all subsequences converge to the same limit, then the sequence diverges.
2. **(Divergence Boundedness Criterion)** If a sequence is unbounded, then it diverges.
3. **(Divergence Description)** If a sequence diverges, then it is not monotone or not bounded.

Focusing on the property of divergence (Div.), let's take a moment to differentiate between the condition and the consequence. Directional statements of the form 'Div. $\implies B$' (with divergence as the condition) provide a *description of* divergent sequences; whereas '$A \implies$ Div.' (with divergence as the consequence) provide a *criterion for* divergent sequences.

The first two contrapositive statements each give a criterion for divergent sequences. The first says that if not all subsequences converge to the same limit— i.e., if there exists two subsequences that converge to different limits or one sequence that does not converge at all—then we can conclude the original sequence is divergent. The second gives a different criterion which says that if the sequence is unbounded then it diverges. Consider the previously discussed sequence

$$(a_n) = \left(0, \tfrac{1}{2}, 0, \tfrac{3}{4}, 0, \tfrac{5}{6}, 0, \ldots\right).$$

How might we justify that it diverges? This sequence is bounded, so the non-boundedness criterion from the second statement does not apply. Recognizing (a_n) is not monotonic suggests possibly using statement (3), but this statement is not a criterion for divergence since divergence appears in the condition of the statement. The first contrapositive statement is the one we want! Observing that $(0, 0, 0, 0, 0 \ldots)$ is a subsequence whose limit is 0, and $\left(\tfrac{1}{2}, \tfrac{3}{4}, \tfrac{5}{6}, \ldots\right)$ is a subsequence whose limit is 1, statement (1) justifies the conclusion that (a_n) diverges.

Because the third contrapositive statement has divergence in the condition, it articulates a description of divergent sequences (*not* a criterion for them). If we know a sequence diverges, the third statement guarantees the sequence will be either non-monotone or unbounded—or both. Consider the divergent sequence

$$(b_n) = \left(0, 1, 2, \tfrac{1}{4}, 0, 1, 2, \tfrac{1}{8}, 0, 1, 2, \tfrac{1}{12}, \dots\right).$$

Because it diverges, (b_n) must be non-monotonic or unbounded. A little inspection reveals that (b_n) is not monotonic. (Other divergent sequences like $(c_n) = (\sqrt{1}, \sqrt{2}, \sqrt{3}, \sqrt{4}, \sqrt{5}, \dots)$ are monotonic, but unbounded.) But we cannot guarantee, for example, that there are no convergent subsequences in (b_n)—$(0, 0, 0, \dots)$ is one; nor can we guarantee there would not even be subsequences that converged to the same limit—$\left(\tfrac{1}{4}, \tfrac{1}{8}, \tfrac{1}{12}, \dots\right)$ also converges to 0. Which is to say the property in statement (1) does not describe something true of all divergent sequences, even though it gives a condition for divergence. If we flip the question around and ask for a proof that (b_n) really diverges, then we would have to turn our attention back to statements (1) and (2). Because it has two subsequences converging to different limits—e.g., $(0, 0, 0, \dots)$ and $(1, 1, 1, \dots)$—(b_n) diverges by the criterion in statement (1).

5.4   Connecting to Secondary Teaching

In the initial teaching situation, the teacher made several statements, each expressing some intended mathematical relationship. By considering students' potential interpretations, we highlighted some of the challenges of classroom communication. When the intended, expressed, and interpreted logical meaning don't all match then confusions ('iff', 'is', 'converse') arise—the proper mathematical relationship can literally be lost in translation.

The teacher is responsible for being attentive to the way everyday communication intersects with mathematical meaning. We have argued that teachers need to use everyday language in class and that they should not always state concepts in formal terms. To be consistent with TP.3, this means paying special attention to logical relationships as they might be interpreted, and not just as they were intended. Informal does not mean imprecise, and the principles of logic can still be invoked in a useful way. Distinguishing between the condition and the consequence of a directional relationship determines whether it communicates a description or a criterion, and switching to the contrapositive also has the potential to add new insight.

5.4.1 Counterexamples

Let's look at these ideas by continuing the teaching scenario from the beginning of the chapter.

Ms. Rojas (who previously defined a trapezoid to be a quadrilateral with exactly one pair of parallel sides) gives statement (1) as a definition: "A trapezoid is isosceles if the non-parallel opposing sides are congruent." After proving that the diagonals of an isosceles trapezoid are congruent, Ms. Rojas makes statement (2):

> This is the same as saying that an isosceles trapezoid is a quadrilateral with congruent diagonals.

Giving this some thought, a student named Adya suggests the statement is incorrect. Adya goes to the board and draws a rectangle which she proposes is a counterexample.

Ms. Rojas first asks Adya for some **content** clarification:

> You drew a rectangle. How are you relating a rectangle and an isosceles trapezoid?

After Adya responds, Ms. Rojas follows up with a **logic** question:

> You drew a quadrilateral with congruent diagonals. Tell me more about why this is a counterexample?

Sometimes confusion is about mathematical content, but the theme of this chapter is that it can also result from a misunderstanding of logical relationships. As a teacher, addressing both *content* and *logic* matters; pursuing one without the other might not get to the heart of the confusion. Both are evident in Ms. Rojas's response.

Ms. Rojas's first content question is motivated by a desire to understand whether Adya thinks a rectangle is a special case of an isosceles trapezoid. The follow-up logic question explores whether she is interpreting the statement to be one of expressing equivalence ($\Longleftrightarrow$) of the two parts, or of expressing a property ($\Longrightarrow$) of isosceles trapezoids. Adya's responses can help identify the actual source of confusion.

Let's look at what this means in terms of two student profiles.

Student Profile 1 A student who thinks a rectangle is an isosceles trapezoid.

Suppose in response to the first content question, the student explains the rectangle by saying it is an example of an isosceles trapezoid. In this case, a teacher can point out that because a trapezoid was defined to have exactly one pair of parallel sides, a rectangle is in fact not a trapezoid at all (and hence not an isosceles trapezoid). This would also be an opportunity to make a point about counterexamples more generally. Since a rectangle has congruent diagonals, it would not qualify as a counterexample. A proper counterexample would require finding a quadrilateral whose diagonals are not congruent that still managed to be an isosceles trapezoid.

Student Profile 2 A student who thinks the statement is bidirectional (because of the "is"), and that a rectangle is not an isosceles trapezoid.

Suppose in response to the second (logic) question, the student explains she is trying to show that a quadrilateral with congruent diagonals and an isosceles trapezoid are not the same thing. In this case, the teacher can start by affirming that, if this were a bidirectional statement, a rectangle would be an excellent counterexample. A rectangle is indeed a quadrilateral with congruent diagonals that is not an isosceles trapezoid. The teacher can then go on to address the 'is' confusion. Acknowledging the potential for misunderstanding, the teacher can clarify that her statement was intended to communicate a conditional relationship—specifically, that isosceles trapezoids have the property of being quadrilaterals with congruent diagonals. The statement was *not* trying to suggest the converse—that all quadrilaterals with congruent diagonals are isosceles trapezoids. It provides a *description of*, not a *criterion for*, isosceles trapezoids.

Even though the proposed counterexample was slightly off the mark in this case, we can still observe the way its use falls under the umbrella of TP.2. Counterexamples are an especially valuable type of special case because they can be used to test and scrutinize mathematical ideas—or, more pointedly, to demonstrate the falseness of a claim with a single example.

5.4.2 Converses

Let's consider a different continuation of the teaching scenario where, this time, the teacher's statement brings up the logical issue of the converse:

> Ms. Rojas gives statement (1) as the definition for isosceles trapezoid and, after proving that the diagonals of an isosceles trapezoid are congruent, she makes statement (3):
>
> This is the same as saying that a trapezoid is isosceles if the diagonals are congruent.
>
> In response, Adya asks the following question:
>
> Are you saying that any trapezoid with congruent diagonals is isosceles?

We look at how this might play out in terms of two teacher profiles.

Teacher Profile 1 The teacher intended to restate that isosceles trapezoids have congruent diagonals.

As teachers, it is important to interrogate our own statements as well as those of our students. Although students are known to reverse the condition and the consequence of a statement, in this case, it is the teacher who has accidentally done the switching. Recognizing the error, the teacher should affirm to the student that, no, this was not what she intended to communicate. In terms of logic, it provides an opportunity to point out the difference between a statement and its converse, and to give examples of each. In this particular case, the class could then investigate whether congruent diagonals imply a trapezoid is isosceles, and then return to statement (3) and rephrase it in more precise language.

Teacher Profile 2 The teacher intended to give a new definition for an isosceles trapezoid.

In this instance, the teacher needs to clarify her use of the phrase, "This is the same as saying…" Logically, the teacher's directional statement is not equivalent to the observation that isosceles trapezoids have congruent diagonals—in fact, it is the converse, and it needs to be independently verified. Once she proves the converse to be true, the teacher can then assert a biconditional relationship in which $A \implies B$ and $B \implies A$. This means congruent diagonals in a trapezoid are a defining feature—a *criterion for*, and a *description of*, every isosceles trapezoid.

5.4.3 Grammatical Variation

As a final point, we consider this discussion in relation to TP.6. Some logical ambiguity is an inevitable and necessary part of teaching. No matter how hard we try, we cannot avoid some measure of confusion that the mixture of formal and informal communication brings to the classroom.

After reading this chapter, you might conclude that the best solution is to avoid semantic ambiguity at all costs—to always give rigorous, logically formulated statements. Under this scenario, teachers should state all conditional claims in the form "if A then B," and all biconditional claims in the form "A if and only if B." A statement such as "A square is a regular quadrilateral" is banned in favor of the more logically clear "A shape is a square if and only if the shape is a regular quadrilateral." The problem, from our point of view, is that teachers must act as a bridge for students' learning; always remaining on the formal mathematical side of the bridge does not work. We would argue for a different strategy. Rather than providing only logically precise statements, teachers should phrase and re-phrase mathematical ideas in multiple ways. We should intentionally *vary* the grammatical structure in order to flesh out the intended mathematical meanings. It is fine to say "A square is a regular quadrilateral" on one occasion if we complement it on other occasions with statements like "All quadrilaterals that are regular are squares", "Being a square implies being a regular quadrilateral, and vice versa," and "The set of squares is the same as the set of regular quadrilaterals." Re-phrasing the relationship in different forms communicates nuances that students might miss from a single formulation. This kind of grammatical variation reinforces TP.6—that a teacher should have multiple ways to explain the same idea—and it has additional relevance in this context. By considering how logic underpins mathematical interpretation, we are inviting students to enter into characteristically mathematical ways of thinking and learning.

Problems

5.1 Consider the statements, "A square is a regular quadrilateral" and "A square is a rectangle." (i) For each, write the correct mathematical relationship as a conditional ($\Longrightarrow$) if-then statement or a biconditional ($\Longleftrightarrow$) if-and-only-if statement. Then write the converse, inverse, and contrapositive statements of any conditional statements. (ii) Next, suppose a student interprets the "is" in a way that is opposite the intended meaning in each statement—the 'is' confusion. Describe how you might respond to a student who is confused about each statement? Make sure to address both content and logic issues in your response.

5.2 Consider the three isosceles trapezoid statements from the teaching situation, re-written to capture the logical structure of a directional relationship:

1. If a trapezoid has opposing sides that are non-parallel and congruent, then it is an isosceles trapezoid
2. If a quadrilateral is an isosceles trapezoid, then it is a quadrilateral with congruent diagonals
3. If a quadrilateral is a trapezoid and has congruent diagonals, then it is an isosceles trapezoid.

Write the contrapositives for each statement. Discuss which contrapositive statements provide a 'description of' non-isosceles-trapezoids, as well as what that description would be; and which provide a 'criterion for' being non-isosceles-trapezoids, as well as what that criterion would be.

5.3 In class, an Algebra II teacher states, "If two functions are inverses of each other, then their graphs are reflections over the line $y = x$." On one of the practice problems in class, a student looks at a graph, and says: "Well, their graphs are reflections over the line $y = x$ so they are inverse functions." First, create a counterexample to the student's claim. Then, explain how you, as the teacher, would respond to the student and why. Make sure to address both content and logic issues in your response.

5.4 Read the short description of the classroom situation below:

Teacher: The slope of the graphed line is 2, which means the coefficient in the equation is 2.

Student: What's a coefficient?

Teacher: A coefficient is the number in front of x in the equation. The coefficient changes the slope of a function.

First, write a description of two different ways that someone might *interpret* the teacher's response to the student. Second, rephrase the above dialogue, as it has been expressed, into a definition for coefficient and an if-and-only-if statement about slopes. Third, decide whether the definition for coefficient as expressed is appropriate, and whether the if-and-only-if statement is true—show some examples you used to help in your decision. If the definition is not appropriate, or the if-and-only-if statement is not true, modify them so they are valid.

5.5 A high school teacher is helping students learn to solve quadratic equations. The example $(x+2)(x+3) = 0$ is on the board. The teacher states, "Well, we know that if either $(x + 2)$ or $(x + 3)$ is zero, then the product will be zero. So, to solve equations like this and find values for x which make the product 0, we write $x+2 = 0$ or $x + 3 = 0$, which gives us $x = -2$ or $x = -3$ as our solutions." Translate the teacher's first sentence and second sentence into formal logical statements (the statements are not false). (Note, the second sentences can be framed in similar terms as the first.) Explain the logical error the teacher has made—that is, why the first sentence does not provide a logical justification for the second sentence about how to solve quadratic equations.

5.6 This problem builds on the previous Problem 5.5. Suppose a high school teacher is helping students learn to solve quadratic inequalities, such as $(x+2)(x+3) > 0$. The following statement is true: "If $(x + 2) > 0$ and $(x + 3) > 0$, then we know $(x+2)(x+3) > 0$." What is the logical error about this statement that would be analogous to the one the teacher made in the previous exercise? Explain why the problem is magnified in this context. Rephrase this statement to be a logically correct and complete statement for solving quadratic inequalities. Describe how this might inform your teaching of solving inequalities to secondary students.

5.7 The ideas of logic underpin the algebraic work of solving equations. For this problem, we offer some explanation before asking you to complete the task.

Explanation When we write an equation with variable expressions, or are given one, we can consider such equations, and the algebraic solving process, as logical statements about solution sets. Here, we consider single-variable equations, because they are very common in secondary mathematics. We can interpret the equation $2x + 1 = 15$ as the statement, "x is a solution to the equation $2x + 1 = 15$". Algebraic solving processes are then logical statements about solution sets: (if possible) we want to write another equation that has the same solution set as the original. Consider the following example:

Equation	Logical statement
(1) $2x + 1 = 15$	"x is a solution to the equation $2x + 1 = 15$"
(2) $2x = 14$	"x is a solution to (1) $\Longleftrightarrow$ x is a solution to (2)"
(3) $x = 7$	"x is a solution to (2) $\Longleftrightarrow$ x is a solution to (3)"

What we find here is that the algebraic steps are connected by if-and-only-if ($\Longleftrightarrow$) statements about the solution sets. This is because adding or subtracting to both sides of an equation, and dividing by a non-zero value to both sides of an equation, are algebraic steps that preserve the solution set—they are some of the axioms of equality. And although we often see this sequence of algebraic steps from top to bottom, i.e., in the order that we write them, what is especially important is the sequence of logical statements from *bottom to top*. This is because we want the end product, statement (3), to tell us something about the initial problem, statement (1), and vice versa. What ends up being important is the logical chain(s) we can form about solutions sets. In this case: (3) $\Longrightarrow$ (2) $\Longrightarrow$ (1), which means "if x is a solution to $x = 7$, then x is a solution to $2x + 1 = 15$"; *and* (1) $\Longrightarrow$ (2) $\Longrightarrow$ (3), which means "if x is a solution to $2x + 1 = 15$, then x is a solution to $x = 7$." This essentially means that solutions to $2x + 1 = 15$ are *identical* to solutions to $x = 7$, which clearly has exactly one solution. However, not all steps in algebraic solution processes are connected by if-and-only-if statements.

Problem Prompt   Consider the following algebraic solution:

Equation	Logical statement
(1) $(x + 3)^2 = 4$	"x is a solution to the equation $(x + 3)^2 = 4$"
(2) $x + 3 = 2$	"x is a solution to (1) $\Longleftarrow$ x is a solution to (2)"
(3) $x = -1$	"x is a solution to (2) $\Longleftrightarrow$ x is a solution to (3)"

Look carefully at the logical statements in the right-hand column. First, explain why step (2) is connected by a directional implication ($\Longleftarrow$) and not if-and-only-if ($\Longleftrightarrow$). Then write the logical chains and conclusions we can make between $(x + 3)^2 = 4$ and $x = -1$ in *both directions*. Do we have all the solutions? Can we have any extraneous solutions? How do you know? What happened?

5.8 This problem builds on ideas from the previous Problem 5.7. For the following algebraic solution, write the corresponding logical statements for each of the algebraic steps, and write an interpretation for what these mean in terms of the solution set. [Note, you should end up discussing the idea of *extraneous solutions*.] Describe how you would respond to a student who asks why, and at what point, the extraneous solution came into the solving process.

Equation	Logical statement
(1) $\sqrt{x+3} = x - 9$	
(2) $x + 3 = (x - 9)^2$	
(3) $x + 3 = x^2 - 18x + 81$	
(4) $0 = x^2 - 19x + 78$	
(5) $0 = (x - 6)(x - 13)$	
(6) $x - 6 = 0$ or $x - 13 = 0$	
(7) $x = 6$ or $x = 13$	

5.9 Two teachers were trying to explain solving systems of linear equations and they said the following:

> Teacher A: All solutions of a system of linear equations can be found via either elimination or substitution.
> Teacher B: If a system of linear equations has one solution, you can find it via elimination or substitution.

(i) Write each statement as an if-then solution. (ii) Describe what the mathematical differences are between these two statements. (iii) Based on the mathematical differences between the statements, state which would be better to tell students and why (you might refer to TP.1 in your response).

5.10 In the language of real analysis, a *sequence* and a *series* are different mathematical objects. A series is an infinite sum of real numbers—that is, it replaces all the commas separating terms in a sequence by addition signs. Yet, in Definition 2.4.3, Abbott first defines an infinite series, and then states: "We define the corresponding *sequence of partial sums* (s_m) by $s_m = b_1 + b_2 + b_3 + \ldots + b_m$, and say that the series converges to B if the sequence (s_m) converges to B" [1, p. 57]. That is, he describes a way to turn a series into a sequence (a previously-studied object). Explain what teaching principle you would say this describes.

References

1. Abbott, S. (2015). *Understanding analysis* (2nd ed.). New York, NY: Springer.
2. Euclid., Heath, T. L., & Densmore, D. (2002). *Euclid's Elements: All thirteen books complete in one volume, the Thomas L. Heath translation*. Santa Fe, NM: Green Lion Press.

Continuity and Definitions

6.1   Statement of the Teaching Problem

Definitions play a fundamental role in mathematics. Because mathematical objects do not exist in a physical sense—they are abstract—definitions are necessary in order for us to have a proper sense of the objects we are studying. In most aspects of life, definitions are *extracted* from a collection of pre-existing examples so that the definition flows from an attempt to describe the objects being defined. Perfect precision is not typically a requirement. Debating whether a hot dog is a sandwich or a stool is a chair illustrates that what constitutes a "sandwich" is pretty flexible and there might not be a definition for "chair" at all. In the words of U.S. Supreme Court Justice Potter Stewart, "I'll know it when I see it" is usually good enough for daily life. Not so in mathematics. While definitions in mathematics are often extracted from a collection of examples, once they have been established they become *stipulative*—the definition precisely bounds and specifies a concept so that any object which meets the defining criteria is considered an example.[1]

Much work goes into the process of crafting definitions. It can be difficult to generate a definition that unambiguously captures a specific set of objects—one that extracts the most salient characteristics and matches our intentions and intuitions. For example, the history of mathematics is full of different kinds of functions (e.g., polynomials, trigonometric functions, logarithms), but it was only in the last century that mathematicians attempted to formulate a proper definition of "function," and there remains a range of options about how the definition should be phrased. (We give a definition for this text in the next chapter.) In some cases, a function is defined to be "a mapping relating a set of input values to a set of output values where each input is related to exactly one output." Another common definition is "a collection of ordered pairs (x, y) where x comes from a set X, y comes from a set Y, and no

---

[1] See Edwards and Ward [2] for further discussion about definitions in mathematics classrooms.

© The Author(s), under exclusive license to Springer Nature Switzerland AG 2022   73
N. H. Wasserman et al., *Understanding Analysis and its Connections to Secondary Mathematics Teaching*, Springer Texts in Education,
https://doi.org/10.1007/978-3-030-89198-5_6

two ordered pairs have the same first coordinate." Is one better? Are they the same? Is either an improvement over the intuitive idea that a function is just a formula that relates *x* to *y*?

Whereas arguing about whether a hot dog is a sandwich is a harmless way to pass the time, proving theorems about functions requires that there be no confusion about what qualifies as a function. Once a definition is agreed upon, it becomes the foundation for mathematical study. Intuition can still be a guide, but any implications or properties that follow must flow logically from the definition.[2] Unambiguous definitions are paramount to the deductive process of mathematics, but they are not set in stone or handed down from on high. Crafting rigorous definitions is a human endeavor and, as such, there is not always agreement on what they should be.

Consider the following pedagogical situation:

Trapezoids have different definitions. Texas uses an *exclusive* definition:

1. A trapezoid is defined as a quadrilateral with *exactly one pair* of parallel sides

New York uses an *inclusive* definition:

2. A trapezoid is defined as a quadrilateral with *at least one pair* of parallel sides

Ms. Abara, a geometry teacher in Texas, has already taught her students about different kinds of quadrilaterals when a new student named Lena arrives from New York. After a few days, Ms. Abara senses that Lena disagrees with the other students about whether or not certain quadrilaterals are trapezoids. Ms. Abara is trying to figure out how best to resolve this issue with Lena, as well as how to talk about isosceles trapezoids with the class. How might she respond?

Having two definitions for a trapezoid is not necessarily problematic. Many concepts are defined in multiple ways, and definitions that appear to be different can sometimes turn out to be logically *equivalent*, meaning they specify the same set of objects. For example, the two definitions for "function" given above are essentially equivalent. The first uses less formal language than the second, but an object deemed a function according to one would also be a function according to the other. Two definitions being equivalent is not necessarily problematic in a classroom. A

---

[2] The relationship between definition and theorem is not quite this uni-directional. The key point is not necessarily about a chronological ordering, but a semantic one—theorems draw on definitions.

textbook typically chooses one as the definition and explains how the other is a consequence of the first, based on how the author wants the content to be structured. However, two definitions are *competing* when they specify different sets of objects for the same concept. This situation poses more of a challenge in teaching. How definitions are stated can potentially clarify or obscure the mathematical objects being studied. Moreover, the definition is the starting point, and so it determines the properties and implications that follow and subsequent definitions all have to be crafted with respect to the original choice.

Before moving on, think about which definition of trapezoid from the pedagogical situation is most familiar. Are these two definitions equivalent or competing? How would you define an isosceles trapezoid based on each definition? Which definition is better, in your opinion, and how does your choice reflect the characteristics you value in a mathematical definition?

6.2 Connecting to Secondary Mathematics

6.2.1 Problematizing Teaching and the Pedagogical Situation

The two definitions for trapezoid are competing definitions because they specify different sets of objects. We elaborate on why and problematize two ways a teacher might respond to the issues in the pedagogical scenario.

One way to approach competing definitions is simply to pick one, assert it as the correct definition, and reject the other. Although this would resolve the tension, it seems odd to say that one state's definition is "incorrect." It would make more sense to say that different state education boards make different choices, and in this class we will use a particular definition. Moreover, declaring that one definition is wrong misrepresents how definitions operate within mathematics. Such a response does not convey the "human construct" nature of definitions. Some definitions are more normative, or standard, and there is a temptation to declare less normative definitions to be incorrect, but at some level they are not wrong, just different. They designate a distinct set of objects for study. Simply opting for one definition over the other misses an opportunity to highlight the stipulative nature of definitions in mathematics. Once a definition is given, the collection of objects characterized by that definition becomes the domain of study. Our personal opinion as to what objects should qualify no longer matters. We have to adapt to consider all possible objects that fulfill the definition, even if the resulting collection is different from what we might have preferred or expected. In the teaching scenario, why the definitions are competing has to do with how trapezoids relate to parallelograms; either *no* parallelograms are trapezoids or *all* parallelograms are trapezoids.

The teacher's response in the pedagogical situation also has implications for how to define the concept of an *isosceles* trapezoid. This is a bit surprising. A first impression is that what makes a trapezoid isosceles should be evident by applying the criteria for isosceles to the objects designated by either definition of trapezoid. The problem is that definitions for subsequent concepts build on earlier definitions,

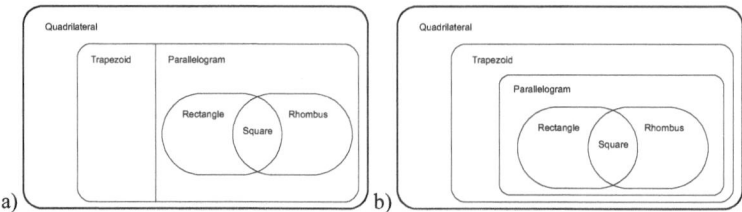

Fig. 6.1 Comparing the (**a**) exclusive and (**b**) inclusive definitions of trapezoid

and this causes some complications. Consider a standard definition for isosceles trapezoid, given in Chap. 5, which asserts that "a trapezoid is isosceles if the non-parallel opposing sides are congruent." This definition assumes the exclusive definition of trapezoid. The criterion for isosceles does not make sense if we have two sets of parallel sides, which is a possibility in the inclusive definition. What do we do if there are no non-parallel sides to consider?

Before reading on, think about how you might resolve the issue of defining an isosceles trapezoid so that it makes sense for both definitions of trapezoid.

6.2.2 Trapezoids

The essential difference between the two definitions is the way trapezoids relate to parallelograms. This is illustrated in the Venn diagram in Fig. 6.1.

The exclusive definition captures the intuitive idea that a trapezoid can be created by slicing off the top of a triangle. By adopting this definition trapezoids are required to have non-parallel sides and so parallelograms are not trapezoids. Categorizing in this way, a quadrilateral with at least one set of parallel sides is either a trapezoid or a parallelogram—it falls into one category or the other, but not both. They are *disjoint*.

If we adopt the inclusive definition, though, trapezoids encompass parallelograms. Parallelograms are a nested *subset* of trapezoids. In this categorization, every parallelogram is simultaneously a trapezoid. Although this categorization might feel unusual at first, it is a familiar way to structure definitions. "All squares are rectangles but not all rectangles are squares" expresses the same type of nested relationship. If we consider number sets, the natural numbers are a subset of the integers, which are a subset of the rationals, and so on. With this nested structure, if a property is true for the objects in a set, then it necessarily applies to a nested subset. As an example, a trapezoid's area is found by the formula $A_{trap} = \frac{1}{2}(b_1 + b_2)h$, where b_1 and b_2 are the lengths of two of its parallel sides and h is the perpendicular height between those sides. With the exclusive definition, it cannot be assumed that this formula will also be true for the area of a parallelogram. But using the inclusive definition, a parallelogram's area must be given by the same formula. With the extra condition $b_1 = b_2 = b$, the formula gives $A_{par} = \frac{1}{2}(2b)h = bh$, illustrating

Fig. 6.2 Defining isosceles trapezoid with the exclusive definition of trapezoid

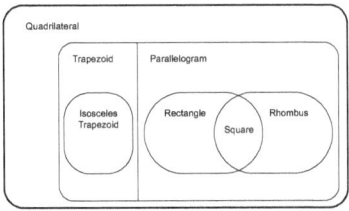

how nested categories can highlight connections between related objects and their properties.

6.2.3 Isosceles Trapezoids

Defining an isosceles trapezoid as one in which "non-parallel opposing sides are congruent" works well with the exclusive definition for trapezoid. It makes isosceles trapezoids a proper subset of trapezoids (Fig. 6.2). Because parallelograms are not trapezoids, there are no questions about whether a parallelogram is an isosceles trapezoid.

This definition does not work well with the inclusive definition of trapezoid; it does not resolve the question of whether a parallelogram is an isosceles trapezoid. We consider two possibilities for defining an isosceles trapezoid in the inclusive case.

As a starting point, we could try to adapt our current criterion for an isosceles trapezoid to work with the exclusive definition. What happens if we drop the "non-parallel" stipulation and rephrase the criterion as: *"A trapezoid is isosceles if it has a pair of opposing sides that are congruent."* For any quadrilateral that qualifies as a trapezoid under the exclusive definition (i.e., a trapezoid that is not a parallelogram), we get the same conclusion we did before. Meaning any trapezoid that was previously classified as isosceles retains that designation with this adapted criteria.[3] So far so good. Now let's see what happens when we move to the inclusive definition and consider a trapezoid that is also a parallelogram. Notably, our amended definition of isosceles is meaningful in this context since we can assign a truth value to the existence of a pair of congruent opposing sides. Because every parallelogram has (two) pairs of congruent opposing sides, all parallelograms are isosceles trapezoids. In terms of our Venn diagram, this means we would add an additional nested set to specify isosceles trapezoids with the inclusive definition (see Fig. 6.3a).

But does this classification of isosceles trapezoids agree with what our intuition tells us to expect? This is hard to say, and there are likely varying opinions on the matter. Extending the notions of congruence and symmetry familiar from isosceles triangles, though, we can make a list of several properties that we might naturally associate with isosceles trapezoids:

---

[3] If it is the parallel sides that were the ones congruent, it would be a parallelogram and hence not a trapezoid.

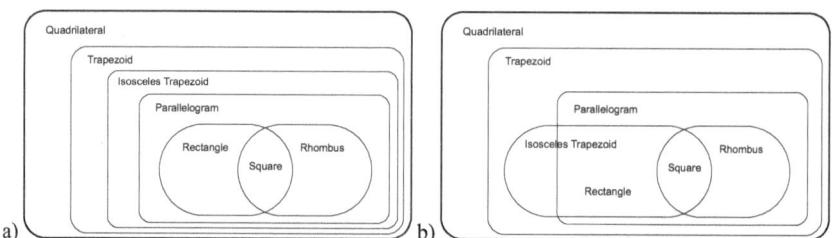

Fig. 6.3 Two approaches to defining isosceles trapezoids, with the inclusive definition of trapezoid

- opposite sides congruent
- base angles congruent (i.e., a pair of consecutive angles that are congruent)
- a line of symmetry
- diagonals congruent

To explore the possible differences in these characterizations, let's fashion another possible definition of isosceles based on the second item in the list: "*A trapezoid is isosceles if it has a pair of consecutive angles (or base angles) that are congruent.*" Again, for any quadrilateral that qualifies as a trapezoid under the exclusive definition (i.e., a trapezoid that is not a parallelogram), we get the same conclusion we did with the original formulation. That is, we could prove a theorem that says, "A trapezoid has a pair of non-parallel sides that are congruent if and only if it has a pair of congruent consecutive angles." But now using the inclusive definition, where parallelograms are trapezoids, we need to consider which types of parallelograms have a pair of congruent consecutive angles. Rectangles and squares have pairs of congruent consecutive angles, but other parallelograms do not. For consecutive angles of a parallelogram to be congruent, the sides need to perpendicular. This approach to defining isosceles trapezoids, which is the more normative approach with an inclusive definition of trapezoid, configures subsets of quadrilaterals quite differently than before (Fig. 6.3b).

What intuition about isosceles trapezoids does this second definition capture? Which definition feels like the "right" one in this inclusive context?

6.3   Connecting to Real Analysis

Continuous functions are a central point of study in an analysis course. Throughout high school and university mathematics, the concept of continuity gets described in ways that range from informal to excessively precise. Here, we consider four definitions of continuity, some of which are likely familiar to you. The initial question is whether the definitions are equivalent or competing. Do these definitions classify the same set of functions as continuous? Be sure to consider atypical functions as you sort through the four proposals.

(Possible) Definition For each proposed definition we consider a real-valued function $f : A \rightarrow \mathbb{R}$, where A is a subset of $\mathbb{R}$.

1. The function f is **continuous on** A if its graph can be drawn without lifting up one's pencil.
2. The function f is **continuous on** A if for every $c \in A$, $\lim_{x \to c} f(x) = f(c)$.
3. The function f is **continuous on** A if for every $c \in A$ and $\varepsilon > 0$, there exists a $\delta > 0$ such that if $|x - c| < \delta$ (and $x \in A$) then $|f(x) - f(c)| < \varepsilon$.
4. The function f is **continuous on** A if $\lim_{n \to \infty} f(x_n) = f(c)$ for every sequence (x_n) (with $x_n \in A$) that converges to some $c \in A$.

Most of us come to the table with a set of expectations for what continuity entails. Polynomials are continuous; so are sine and cosine curves. Continuous functions should not have holes or jumps. Which of these definitions capture our sense of what continuity should be? Which functions are included and which are ruled out? Are there stipulations in some of these definitions that might surprise us or push against our intuition? Remembering the pitfalls we experienced defining isosceles trapezoids, are there situations where the definitions don't make sense? Is the given criteria precise enough that it can be evaluated at all?

6.3.1 Considering Various Definitions of Continuity

To get a better sense of each of these four proposed definitions for continuity, let's try them out on some example functions.

Example Consider the function f depicted in the graph

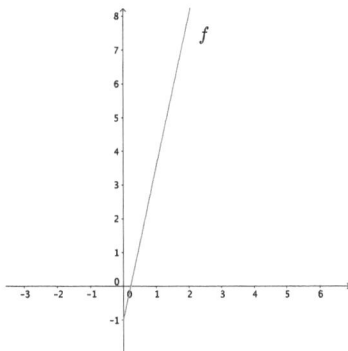

Tasked with deciding whether f is continuous from just this graphical information, Definition (1) feels like an appropriate and straightforward way to proceed. The graph can be drawn without lifting up one's pencil, so f appears to be continuous. But how compelling is this argument?

Fig. 6.4 Zoomed out graph
of f, discontinuous by
Definition (1)

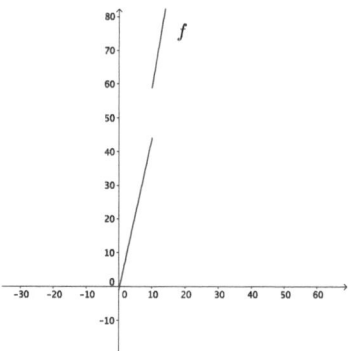

This example illustrates a weakness in the first definition. Relying on a graph
for our definition of continuity is inherently limiting. There are functions that
cannot be graphed in a meaningful way. One example would be the discontinuous
Dirichlet function, $g(x) = \begin{cases} 1 : x \in \mathbb{Q} \\ 0 : x \notin \mathbb{Q} \end{cases}$. How can we apply Definition (1) when
g cannot be graphed? An example of an ungraphable function that is continuous
is the Weierstrass Function. Its progressively finer layers of oscillations outstrip
the resolution of any graphing device (the function has a self-replicating, fractal,
nature). Generally speaking, graphs are visual summaries—useful for our intuition
but, by their nature, incomplete. Unless the domain of a function is a finite set of
points, a graph can only provide a partial description. Returning to the function in
the graph above, suppose $f(x) = 1.5x \cdot \lfloor 0.1x + 3 \rfloor - 1$ (for $x \in \mathbb{R}^+$). Indeed,
this is what generated the graph of f. However, now look at a plot of this function,
zoomed out, in Fig. 6.4. By expanding the viewing window, we see that f has a
"jump." This leads to the conclusion that f is discontinuous because we have to
pick up our pencil to draw it. The moral of this story is that Definition (1) has
some severe limitations—criteria based on a function's graph is difficult to evaluate
consistently.

Example Consider the function $g(x) = \frac{1}{x}$ depicted in the graph

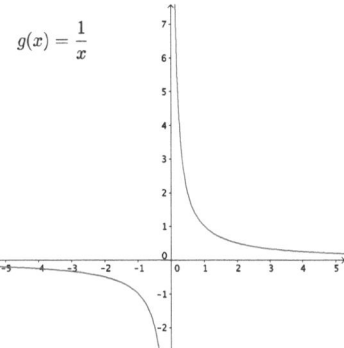

$$g(x) = \frac{1}{x}$$

By Definition (1), $g(x)$ would be discontinuous because drawing the graph requires us to lift our pencil. The problem occurs at $x = 0$, but $x = 0$ is not even in the domain of g! In this example, Definition (1) holds g accountable for its behavior at a point where g is not even defined. Notably, the continuity of a function is typically defined with respect to a specified domain A. In the previous example, f had domain $\mathbb{R}^+$. Although we judged f to be discontinuous, this was due to the jump in the graph and not because it happened not to be defined for $x \leq 0$.

This points to an interesting quality, and perhaps a counterintuitive implication, of Definitions (2), (3), and (4). Each of these latter three definitions explicitly requires us to investigate the behavior of the function in question at individual points $c \in A$. Each definition is imbued with a particular way to define continuity at a point, and defines a function to be "continuous on A" if it is continuous at each point of A. The definitions do not consider what happens at points outside the intended domain. For $g(x) = 1/x$, the natural choice for the domain is $A = \{x \in \mathbb{R} : x \neq 0\}$, and by Definitions (2), (3), and (4), it turns out that g is indeed continuous on A. If we choose an arbitrary $c \neq 0$, the specified criteria in (2), (3), or (4) is met at c and therefore the function is continuous. (This requires some thought and you are encouraged to pause and think about why this is true in each case.)

It may feel a bit strange to assert that $g(x) = 1/x$ is continuous when it just looks so discontinuous. This is an example of what it means to adopt a formal definition and then live by all its stipulations. The feeling that g is not continuous, which arises from a natural sympathy for the sentiments in Definition (1), must be set aside in favor of the desire to structure our theory of continuity in a rigorous way. That said, it is still the case that some calculus books refer to g as having an "infinite discontinuity" at $x = 0$. One way to make that statement align with our formal definitions is to add $c = 0$ to the domain of g. For instance, we could set $g(0) = 0$ so that A is now all of $\mathbb{R}$. Setting $c = 0$ in Definition (2), we can observe $g(0) \neq \lim_{x \to 0} g(x)$ because the limit does not exist. This implies g is no longer continuous. Switching to the criteria in Definition (4) yields the same conclusion. Figure 6.5 depicts two sequences in the domain of g that both approach $c = 0$: (x_n) from the left and (z_n) from the right. (Those sequences are depicted on the x-axis.) However, the associated sequences $g(x_n)$ and $g(z_n)$ (depicted on the y-axis) diverge to infinity

Fig. 6.5 The new function g
is not discontinuous at $x = 0$
by Definition (4)

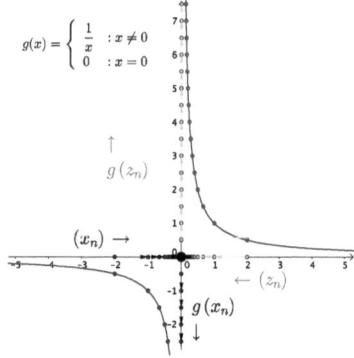

rather than approach $g(0) = 0$. By Definition (4), g is no longer continuous when
its domain is expanded to include 0. (Definition (3) results in the same conclusion
as well.)

Example As a final example, consider the function

$$q(x) = \begin{cases} \frac{1}{2}|x - 1| + 1 & : x \geq 0 \\ 1 & : x = -2 \end{cases}$$

which has the following graph:

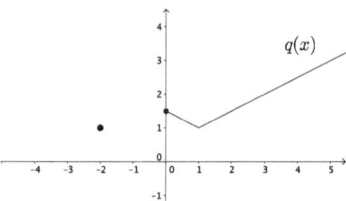

For $q(x)$, we will consider continuity at two particular points of the domain. The
first is $c = 0$, which is an *endpoint* of part of the domain. To use Definition (2) we
need to consider the functional limit as x approaches 0. We know $q(0) = 1.5$, so
the question is whether $\lim_{x \to 0} q(x) = 1.5$. The answer hinges on the definition of
functional limit, and in particular whether it exists at an endpoint like $c = 0$. Using
the definition in Abbott [1] (4.2.1), we can confirm that everything checks out—the
functional limit exists and q is continuous at 0. In fact, q is continuous on the set
$A = \{x \in \mathbb{R} : x \geq 0\}$.

But what happens at the point $c = -2$? This is an *isolated point* of the domain of
$q(x)$, and Abbott's definition stipulates that functional limits can only be considered
at *limit points*, which $c = -2$ is not. This is significant. If we adopt Definition

Fig. 6.6 The function q is
continuous at the isolated
point $c = -2$ using
Definition (3)

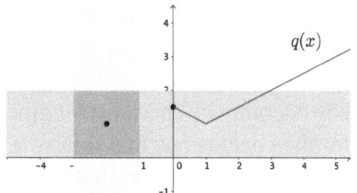

(2), then continuity requires that $\lim_{x \to -2} q(x) = q(-2)$ which is not true. The functional limit is not defined and so using Definition (2) we get that q is *not* continuous on this larger domain that includes the isolated point. This conclusion is different from the one that emerges from adopting either Definition (3) or (4). The logical structure of these latter two statements implies that functions *are* continuous at isolated points of their domain. Definition (2) is thus a competing definition, specifying a different set of functions to be continuous than from Definitions (3) and (4), which turn out to be equivalent to each other.

To understand why these definitions are competing, look carefully at the wording of Definition (3) as it relates to the isolated point $c = -2$, and note especially the parenthetical reminder $(x \in A)$. Given an arbitrary $\varepsilon > 0$, the definition requires us to find a δ neighborhood centered at -2 such that all the points in this neighborhood *that are also in the domain* have y-values within ε of $q(-2) = 1$. Choosing $\delta = 1$ results in the neighborhood $(-3, -1)$, and the only domain point contained in this interval is $c = -2$. With no other x-values to worry about, we conclude that q is continuous. (See Fig. 6.6.) Take a moment to confirm that Definition (4) is structured in a similar way so that a function is determined to be continuous at any isolated points of its domain.

6.3.2 Choosing a Definition

Exploring the examples in this section has revealed some of the strengths and weakness, as well as the logical distinctions, that exist among our four proposed definitions for continuity. From this more informed point of view, how might we settle on the best choice to be our official definition?

Definition (1), although intuitively helpful, must be ruled out on the grounds that it is simply too informal. A proper definition should precisely delineate a set of mathematical objects, and it is not at all clear how to apply the statement in Definition (1) to numerous functions we would like to categorize. As the first two examples show, it can also lead to categorizations that are potentially contradictory.

Definition (2) is appropriately formal, but it categorizes any function with an isolated point in its domain as discontinuous. This competes with Definition (3) (and (4) as well) which is crafted so that isolated points turn out to be points of continuity. How should we decide between these two options? On the one hand, we might feel that isolated points don't innately feel "continuous." The graph of a function defined on the positive integers would amount to a sequence of disconnected dots which is certainly a far cry from a graph that can be drawn without picking up a pencil. This is a reasonable argument for adopting Definition (2), but there are other considerations. The most significant is the way the chosen definition sets the stage for the conclusions that follow. The study of continuous functions is connected to a network of other ideas that are articulated in the theorems of analysis. Details aside, many elegantly stated results such as "continuous functions restricted to compact sets are uniformly continuous," would become laden with awkward disclaimers if we adopted Definition (2). Although it initially pushes against our intuition, classifying functions to be continuous at isolated points turns out to be the more organic way to build the larger theory.

This brings us to Definitions (3) and (4), which we've discussed are equivalent (cf., Abbott's proof of Theorem 4.3.2). Precisely the same set of functions meet their respective criteria for continuity, and both criteria are used widely throughout a course in analysis. This suggests we have a genuine choice to make; *either* could serve as the definition for continuity and then we could prove the other as a theorem to be used as needed. This choice between equivalent definitions means we could use whichever one we found to be most appropriate for the course, the students, etc. While this is true—and there are analysis textbooks that take both approaches—the fact that two statements are logically equivalent does not mean they are equivalent in every respect. There are other considerations, too. When building a mathematical theory, there is an implied hierarchy between a definition and a theorem—definitions are the more primitive foundation on which theorems are built. The distinction is sometimes more art than science and making this distinction usually requires looking forward to see what lies ahead. In the case of continuity, for example, the $\varepsilon - \delta$ criterion in Definition (3) is most amenable to defining the concept of "uniform continuity" referenced in the above result.

Settling on the right definition can be a deliberative and subtle process, but a good sign that you are heading in the right direction is when there is an organic— we might even say poetic—connection between the definition and the theorems that follow. As the mathematician G.H. Hardy famously said, "Beauty is the first test!"

6.4   Connecting to Secondary Teaching

The takeaway from our discussion thus far is that, as teachers, we need to be purposeful about the definitions and explanations we use with students. We have a choice, and that choice has implications for the theorems and definitions that follow. Returning to the competing trapezoid definitions from the beginning of the chapter, the one we choose determines whether parallelograms are distinct from trapezoids or whether they form a nested subset. This arrangement then leads to consequences of its own, shaping the trajectory for the class.

Of our six teaching principles, TP.2 is the most pertinent to this discussion. Whether it's trapezoids, continuity, or some other concept, creating a range of special cases is the best way to probe a proposed definition. This means constructing examples that meet the criteria in the definition as well as those that don't. It also means creating examples that fall near the boundary, barely qualifying as one of the defined objects or falling just shy. These kinds of "near" or "minimal" examples are especially valuable for marking out the scope of a proposed definition and getting a head start on determining the kinds of theorems that it engenders.

6.4.1 Defining Isosceles Trapezoids

Two definitions for the same concept are logically different (i.e., competing) if they specify distinct sets of objects. While many objects will meet the criteria of both definitions, objects that meet one criteria but not the other are the important cases that separate the two definitions. For the two competing ways to define a trapezoid—the exclusive and inclusive definitions—so-called "common" trapezoids with exactly one pair of parallel sides fit both definitions while the set of parallelograms satisfies only the inclusive definition. Illustrating the impact of TP.2, these examples shape our understanding of the tension between the competing definitions and should be at the forefront when we consider the best way to extend the theory of trapezoids with additional definitions and theorems.

Consider the following continuation of the previous teaching scenario:

Ms. Abara gathers the class together and reviews the definition of trapezoid:

In our class, we have defined a trapezoid as a quadrilateral with *exactly one pair* of parallel sides.

She continues by drawing attention to parallelograms and rectangles to illustrate the key difference:

What this means is that trapezoids and parallelograms are separate. Parallelograms, including rectangles, are not trapezoids according to the definition we are using.

She then introduces a new definition for isosceles trapezoids:

Isosceles trapezoids are essentially about symmetry. With our definition of trapezoid, we could define them in terms of the non-parallel opposing sides being congruent. But for class purposes, we will define them in terms of another symmetry: a trapezoid is isosceles if the base angles are congruent.

Recognizing that her new student from New York was introduced to a competing definition of trapezoid, Ms. Abara clarifies the definition used in her Texas-based class and then gives an example to shed light on the difference. To define isosceles trapezoids, the teacher focuses on "symmetry" as the essential feature and states that various definitions based on symmetry might be possible. Ultimately, Ms. Abara defines an isosceles trapezoid in a manner that makes sense with either definition of trapezoid. Although a system of definitions can build on one another, using a definition that works equally well across multiple definitions is a worthy consideration. Ms. Abara's definition using congruence of base angles, rather than congruence of two non-parallel opposing sides, is meaningful and effective for either the inclusive or the exclusive definition of trapezoid. A related version of this type of consideration that happens in a geometric context is whether a definition in Euclidean geometry still makes sense in a non-Euclidean setting. Problem 6.8 asks you to think about this issue.

The process of defining terms and generating examples is an important component of students' mathematical education. Students should also *experience* this aspect of mathematics in an active way. As teachers, it is also important to engage students in this process, recognizing them as independent thinkers capable of refining their own definitions and appreciating the objects their definitions describe.

6.4.2 The Relationship Between Definitions and Theorems

If we choose the definition of isosceles trapezoid to be a trapezoid with congruent base angles, the next logical step is to prove a theorem which states that the opposite sides of an isosceles trapezoid are congruent as well. The definitions established by the teacher lay out the logical trajectory for the class, which can have implications for how students view the larger theory.

To appreciate how the definition-theorem relationship can impact student understanding, consider two possible ways we might choose to define a rectangle. A first definition could be: "A rectangle is an equiangular quadrilateral." It is important to distinguish between what this definition explicitly *assumes* about rectangles and what it logically *implies*. In this case, a rectangle is assumed to be a quadrilateral with four congruent angles. This definition embeds rectangles as a subset of quadrilaterals, but there is no mention yet of right angles or parallel sides. The fact that all rectangles turn out to be parallelograms is a theorem that has to be proved from the definition.[4]

A second definition could be: "A rectangle is a parallelogram with one right angle." Defining a rectangle in this way means we are nesting rectangles as a subset of parallelograms, which is itself a subset of quadrilaterals. This is conceptually

---

[4] 'Equiang. Quad. $\implies$ Par.': Through the construction of a diagonal, the four angles sum to the interior angles of two triangles. In Euclidean geometry, this sum is two straight angles; a fourth of two straight angles is a right angle, and so each angle is a right angle. This makes the same side interior angles supplementary, which means both pairs of opposite sides are parallel. Hence, a rectangle is a parallelogram—and with at least one right angle.

different from from where we started before. In this case, being equiangular is not part of the definition for rectangle but becomes a theorem we can prove.[5]

These two definitions are not competing—we get the same set of rectangles with either one. So is one better than the other? Which criterion feels more fundamental to the nature of rectangles? Or perhaps we should reject them both in favor of asserting "a rectangle is a quadrilateral with four right angles." This latter statement is not as primitive as either of the other two proposed definitions—it assumes more than is necessary to specify the same set of mathematical objects—but there is an argument that this is what a rectangle really is. Without resolving this debate here, we note that the proofs showing these three definitions are logically equivalent utilize ideas specific to Euclidean geometry. In a non-Euclidean context the definitions can lose their equivalence and start to compete, raising the stakes considerably for deciding which one ought to be the definition of a rectangle to begin with.

For many secondary topics there are a variety of definitions that can be chosen. As teachers, it is important to think through these choices and the implications for how the ideas would then progress. There are significant ramifications for students that result from the different ways teachers sequence the definitions with the theorems and properties that follow.

Problems

6.1 In the previous Chap. 5, we looked at several isosceles trapezoid statements. There, we were assuming the *exclusive* definition of trapezoid. This chapter introduced the *inclusive* definition. Look at several statements or theorems about trapezoids from geometry. Determine the truth value of each, depending on which of the two definitions of trapezoid is used. If possible, give two example theorems that would be true under one definition but not true under the other.

6.2 Zero is an even number. However, students often suggest that zero is neither even nor odd. Which of the following would still be true if all other integers (positive and negative) except zero retain their even or odd status? Justify your response for each statement.

1. even + even = even
2. odd + odd = even
3. even + odd = odd
4. even × even = even
5. odd × odd = odd
6. even × odd = even

---

[5] 'Par. One Rt. Angle $\implies$ Equiang. Quad.': Because lines are parallel in a rectangle, and in Euclidean geometry the same side interior angles are supplementary, this means another angle is a right angle. By repetition, we can conclude the rectangle has four right angles, and so equiangular.

6.3 In geometry, the distance between a line and a point not on the line is defined as the distance along a *perpendicular* line. In statistics, the distance between a line (of best fit) and a point (not on the line) is defined as the distance along a *vertical* line. Are these two definitions equivalent or competing definitions? If they are equivalent, provide a justification. If they are competing, provide an example where they would be different and, if possible, one where they would be the same. Then, discuss why geometry and statistics might define the "distance" between two such points in the way they do.

6.4 A class is asked to prove the following definitions of rectangle are equivalent:

1. A quadrilateral is a rectangle if it is a parallelogram with four right angles.
2. A quadrilateral is a rectangle if it is a parallelogram with one right angle.
3. A quadrilateral is a rectangle if it is a quadrilateral with four right angles.

One student submits: "Assume we have a quadrilateral $ABCD$ that is a parallelogram with four right angles. If we accept Definition (1), then we call it a rectangle. But, obviously, if it has four right angles then it has one right angle, so it also fulfills Definition (2). In addition, all parallelograms are quadrilaterals, so it also fulfills Definition (3). Also, we know that adjacent angles of a parallelogram are supplementary, meaning if there is one right angle in a parallelogram, then we actually know that all four are right." Respond to the following: (i) as the teacher, how would you respond to the student's written work?; (ii) discuss what, if any, errors are present, and what, based on what the student has submitted, would still be needed to complete the question.

6.5 Consider teaching a course in geometry. Give two different ways to structure a sequence of definitions for special quadrilaterals: trapezoids, parallelograms, rectangles, rhombuses, and kites. Give both a precise definition for each, as well as the sequential order you would discuss them with students. Provide a justification for each of the two possible approaches. Then, consider having to teach about area: Which sequence of definitions do you think would be better, or worse, for teaching students about area formulas for quadrilaterals? Explain your reasoning.

6.6 A common definition for the absolute value of a number is piece-wise: $|x| = x$ if $x \geq 0$ and $|x| = -x$ if $x < 0$. Think about whether this definition makes sense for numbers, x, that are Natural numbers? Integers? Rational numbers? Real numbers? Complex numbers? If the definition does not make sense for a number set, explain why not. Now think about an alternate definition: $|x|$ is the distance (measured in the typical way) from the "origin" (0 on a number line, $(0, 0)$ in the plane, etc.). For which number sets would this definition make sense? Discuss which definition you might use with a class of students? Why?

6.7 Think about how you would define the "perimeter" of a (2D) shape (an idea we pick up on in Chap. 10). Compare and contrast the following two possible definitions: (i) The perimeter of a shape is the sum of all the side lengths (on the edge that encloses it); (ii) The perimeter of a shape is the distance around the edge that encloses it. Draw several different kinds of shapes studied in secondary mathematics. Which definition would you use to "define" perimeter? Why? Even though only one is being used as the definition, would the other description of perimeter be discussed with students in any way? If so, how and when might it be discussed?

6.8 In Euclidean geometry, we can define a rectangle in a variety of equivalent ways. Consider the three possibilities below. Which of these definitions makes the most sense in a non-Euclidean geometry context? (In non-Euclidean geometry, the interior angle sum of a triangle does not have to be 180°. Also, there might be multiple lines through a point all parallel to a given line, or there could be no parallel lines through this point that are parallel to the given line.) Explain your reasoning.

- A rectangle is a parallelogram with one right angle
- A rectangle is a quadrilateral with three right angles
- A rectangle is an equiangular quadrilateral

6.9 Consider the following two definitions for a real-valued function f defined on domain A.

- An *increasing function* $f : A \to \mathbb{R}$ is a function such that for $x_1, x_2 \in A$ with $x_1 < x_2$, $f(x_1) \le f(x_2)$.
- A *decreasing function* $f : A \to \mathbb{R}$ is a function such that for $x_1, x_2 \in A$ with $x_1 < x_2$, $f(x_1) \ge f(x_2)$.

If you can, sketch a function that is increasing but not decreasing. One that is decreasing but not increasing. One that is both increasing and decreasing. One that is neither increasing nor decreasing. Explain why your functions meet the required specifications.

6.10 Consider Abbott's Exercise 4.2.10, about the use of left- and right-hand limits in introductory calculus:

Introductory calculus courses typically refer to the right-hand limit of a function as the limit obtained by "letting x approach a from the right-hand side."

(a) Give a proper definition in the style of Definition 4.2.1 for the right-hand and left-hand limit statements:

$$\lim_{x \to a^+} f(x) = L \text{ and } \lim_{x \to a^-} f(x) = L$$

(b) Prove that $\lim_{x \to a} f(x) = L$ if and only if both the right and left-hand limits equal L.

What purpose does part (b) of that exercise serve in terms of a calculus teacher's ability to only discuss functional limits as being in terms of right-hand and left-hand limits? What teaching principle would you consider this example as illustrating?

Turning the Tables

Reflecting on *teaching* from your *learning* in real analysis: TP.4

As another way of connecting to issues of teaching and learning, the "Turning the Tables" sections scattered throughout the text provide additional commentary on some of the ways our teaching principles are exemplified in learning real analysis. Here, we consider TP.4: modeling more complex objects with simpler ones.

Real analysis has many excellent examples of this principle. In Chap. 3, we constructed sequences of rational numbers that converged to a real number. In that example, real numbers with infinite and irregular decimal expansions are being modeled, or approximated, by simpler rational numbers whose decimal expansions terminate. In Chap. 9 we will discuss how the derivative conceptualizes tangent lines as a sequence of secant lines, and in Chap. 12 we'll see how the Riemann integral models the complex region under a curve with a collection of simpler rectangles. These examples collectively illustrate how complex mathematical theories such as a calculus are constructed out of simpler building blocks and make a compelling case for how TP.4 can inform our approach to teaching.

The nested definitions discussed in the present chapter reinforce the central role of TP.4 in the way mathematics is structured. Here, we consider a particular definition from real analysis. The sequential criterion for continuity—given earlier as Definition (4)—states that a function $f : A \to \mathbb{R}$ is continuous at a point $c \in A$ if, for every sequence (x_n) in A converging to c, it follows that $f(x_n)$ converges to $f(c)$. In this characterization, continuity is being defined in terms of *sequences* and *limits of sequences*—both concepts that were previously defined. In this manner, the concept of continuous functions is building on prior ideas. In the spirit of TP.4, we are conceptualizing continuous functions—something relatively complex—by using the simpler device of convergent sequences.

Constructing new concepts and definitions from previously-defined ones is fundamental to how mathematics is organized. Although the primary and more practical significance of TP.4 is best realized in specific examples such as approximating real numbers with rational sequences, the hierarchical nesting of concepts that characterizes mathematics showcases the broad relevance of TP.4. Explicitly making connections to previous concepts as we introduce new ones is a staple of good teaching and another point of connection to TP.4 in the classroom.

References

1. Abbott, S. (2015). *Understanding analysis* (2nd ed.). New York, NY: Springer.
2. Edwards, B. S., & Ward, M. B. (2004). Surprises from mathematics education research: Student (mis) use of mathematical definitions. *The American Mathematical Monthly, 111*(5), 411–424.

The Intermediate Value Theorem and Implicit Assumptions

7.1 Statement of the Teaching Problem

Effective communication, in the classroom and elsewhere, relies on shared under-standing. Perhaps the simplest illustration is with vocabulary. When most people hear the word "plane" they picture a machine that flies through the air. Yet, if a student draws on that understanding while their geometry teacher discusses a "plane," very little that is said will be meaningful to the student. This so-called semantic contamination occurs when everyday definitions interfere with properly understanding mathematical definitions. While the confusion between airplanes and Cartesian planes is not hard to remedy, there are more nuanced levels of semantic contamination that can be amplified in the classroom and are more challenging to address.

When we communicate about mathematical ideas, there is a dichotomy between the rigorous language of formal definitions and the more intuitive type of discourse that focuses on big picture properties and phenomena.[1] Proving theorems requires careful attention to formal definitions, but when most of us discuss a mathematical concept like "function" or "continuity," we are likely referring to the informal collection of examples and ideas that are common from our experience. We may not be thinking about the formal definition. This is especially true for students, and the result is a type of semantic contamination that can go undiagnosed. Students frequently make assumptions about mathematical concepts that stem from a mental image that has built up over time. These assumptions tend to remain *implicit* in communication; it's hard to be explicit about assumptions of which you are not aware.

---

[1] Tall and Vinner [5] make a similar differentiation. They contrast a *concept definition*, by which they mean a formal definition, with a *concept image*, by which they mean all the other interesting things associated with that concept.

© The Author(s), under exclusive license to Springer Nature Switzerland AG 2022 93
N. H. Wasserman et al., *Understanding Analysis and its Connections to Secondary Mathematics Teaching*, Springer Texts in Education,
https://doi.org/10.1007/978-3-030-89198-5_7

Consider the following pedagogical situation:

Mr. Cai, a high school teacher, is illustrating the Intermediate Value Theorem by locating zeros of functions. Using the example $g(x) = x^3 - 3x^2 - 2x + 7$, Mr. Cai points out that since $g(2) = -1$ and $g(3) = 1$, the function must have at least one zero between 2 and 3.

Later, as an "exit ticket," students are asked for a short summary of the key ideas. One student, Chrissy, submits the following response:

> Okay, so if a function is less than 0 somewhere and greater than 0 somewhere else then we know there will be a zero somewhere between them. So in general, if $f(a) < 0$ and $f(b) > 0$, then there is at least one zero in the interval (a, b).

Chrissy's response in the exit ticket appears to be a positive reflection on the lesson—she was able to capture some of the most salient ideas about the theorem. On the other hand, assumptions left implicit can lead to problematic understandings about the mathematics; TP.1 insists these be explicitly acknowledged and revisited. One of the challenges of teaching is the ability to listen to students, interpret their statements, hear potentially implicit assumptions, probe those assumptions, and identify how to respond in order to further mathematical understanding.

Before moving on, think about how you, as a teacher, might respond to the student. What comments might you make? What questions might you ask?

7.2   Connecting to Secondary Mathematics

7.2.1   Problematizing Teaching and the Pedagogical Situation

We problematize some potential responses to the student's summary of the Intermediate Value Theorem (IVT).

A first reaction may be to commend the student. Overall, Chrissy has done a good job summarizing key parts of the IVT. Several aspects are particularly noteworthy. She has successfully generalized the essential ideas from the particular example. This is especially evident from her use of symbolism, such as using '$f(a) < 0$,' and the interval '(a, b).' The student has also noted the importance of the phrase "at least one" zero and made sure to include it in her summary. This is a critical nuance of the IVT. When $f(a) < 0$ and $f(b) > 0$, there could be multiple zeros in (a, b), but we cannot be sure, so "at least one" is the most accurate claim. The student's attention to these details suggests she understands some important ideas about finding zeros and merits a degree of validation from the teacher. But are there other aspects that should be considered? Are there any implicit assumptions or mathematical limitations in the student's statement that, according to TP.1, might need to be clarified?

In response to these questions, a second reaction could involve pointing out that a completely correct answer needs to acknowledge the role of continuity. A function must be continuous on the given interval to apply the IVT. A step function, for example, could be less than 0 at one point, greater than 0 at another, and not have any zeros in between because it "jumps" over the x-axis. In addition to affirmation, a teacher should point out that the function has to be continuous on the interval (a, b) to ensure there is a zero in that interval. Emphasizing this condition to the whole class might also be worthwhile.

Having raised the issue of continuity, the teacher's next job is to ferret out the reason for the omission. What led to this missing component in Chrissy's exit ticket response? Perhaps she is not aware of the significant role of continuity in the IVT, or maybe it was just a careless mistake. Falling somewhere between these two scenarios is the possibility of a subtle form of semantic contamination in Chrissy's use of the term "function." She might be using this term to reference something different than what the teacher, or you as a reader, imagine. Based on the examples she has seen, Chrissy may have been using "function" to mean "polynomials"; or perhaps her mental picture of function only includes continuous ones and so the term was implicitly referring to "continuous functions." What students say and write does not always align with what they understand. Interrogating students' mistakes to unearth their thinking requires asking, "What might the student be assuming in order for this statement to be logical?" The answer frequently includes the existence of implicit assumptions.

Praising Chrissy's response may let a misunderstanding linger, and correcting the response might not address the right misconception. Indeed, one of the things we will see is that, in addition to continuity, there are other assumptions being made about the IVT that are implicit in the student's statement.

7.2.2   Defining Function

Before we consider the IVT further, let's directly address any confusion about the term "function." Based on your own experiences, you probably have a particular image for the concept of function. Before we present a formal definition, pause to consider the different informal images you have for this concept. What do you think about when you consider the notion of a function? What examples? What properties? What pictures or words?

Definition A **function** f is a set of ordered pairs (x, y) such that each x is associated with a unique y. Specifically, $f = \{(x, y)|(x, y_1), (x, y_2) \in f \text{ implies } y_1 = y_2\}$. In this case, we write $f(x) = y$. The set of all x-values is the *domain*, $A = \{x|(x, y) \in f\}$, and the set of all y-values is the *range*, $f(A) = \{y|(x, y) \in f\}$. Any superset $B \supseteq f(A)$ can be the *codomain*, and we write $f : A \to B$.

The definition above is similar to other definitions of function (e.g., Abbott's [1] Definition 1.2.3). The domain A and the co-domain B do not have to be sets of real numbers. The domain could be the set of students in a class and B the set of desks. A function f could be the set of ordered pairs (x, y) where student x sits in desk y.

With a set A (the domain) and a set B (a superset of the range) we could consider different possible collections of ordered pairs (x, y) where $x \in A$ and $y \in B$. The full collection, known as the Cartesian product $A \times B$, contains all possible ordered pairs. Relative to these different possible collections, functions are a particular kind of collection—one in which each x is associated with only one y. This property is known as *univalence*. A second property associated with functions is *totality*, which means every $x \in A$ appears as the first coordinate at least once. In our previous example, totality means that every student gets a desk; univalance means that no student gets more than one. (It's entirely possible, however, that two students share the same desk.) Using this notation, another way to characterize a function is as a subset of ordered pairs from $A \times B$ that is univalent and total on A.

Real analysis and much of secondary mathematics is focused on functions with domains and ranges from the set of real numbers. The graph of such a function on the Cartesian plane $\mathbb{R} \times \mathbb{R}$ is a useful representation of its set of ordered pairs. In fact, you probably imagined particular graphs of functions as you thought about the concept earlier. The formal definition of function is quite broad. Even when we restrict our attention to real-valued functions, there is a wide array of surprising examples that fulfill the defining criteria. Figure 7.1 depicts a range of examples that meet the formal definition. Although you might be tempted to reject them based on your preconceived notion of what a function should look like, the definition demands they become part of your example space. Claims about functions must hold across all possible examples, or be appropriately amended to apply to a particular subclass.

7.3   Connecting to Real Analysis

As we discussed in Chap. 5, conditional statements $(A \implies B)$ have a condition A and a consequence B; if condition A is met we necessarily have B as a consequence. Conditions are the explicit assumptions required for a proposition to hold, but precisely determining the conditions of a theorem can be a bit challenging. Mathematics is a dense language where a lot can be conveyed in a few words and symbols. Fully comprehending all the assumptions of a theorem requires careful scrutiny of the notation to appreciate what is being articulated. It also involves following through to see how the various assumptions are incorporated in the proof. To illustrate the different steps and potential pitfalls in this process, let's take a detailed look at the Intermediate Value Theorem and a corollary we call the Intermediate Zero Theorem.

Theorem (Intermediate Value Theorem) Let $f : [a, b] \to \mathbb{R}$ be continuous. If L is a real number satisfying $f(a) < L < f(b)$ or $f(a) > L > f(b)$, then there exists a $c \in (a, b)$ where $f(c) = L$.

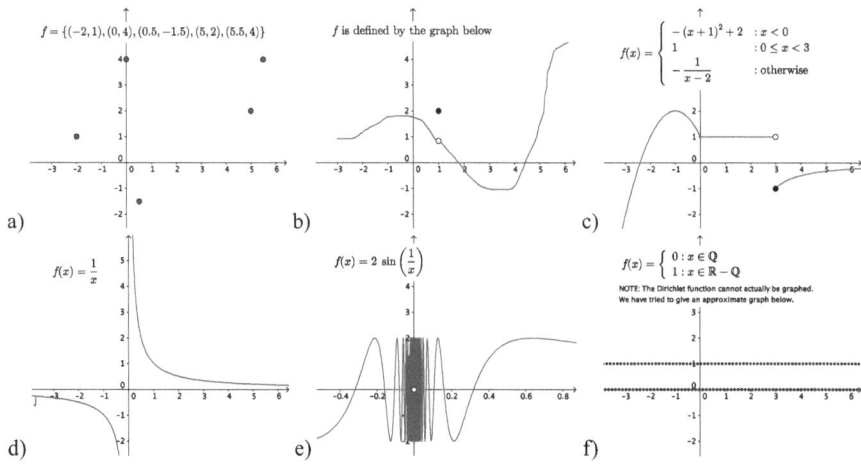

Fig. 7.1 Functions that are: (**a**) sets of discrete points; (**b**) defined by their graphs; (**c**) piecewise defined, and with "jumps"; (**d**) missing domain values, with vertical asymptotes; (**e**) missing domain values, with holes, and arbitrarily close oscillations; and (**f**) defined but cannot be graphed

Theorem (Intermediate Zero Theorem) Let $f : [a, b] \to \mathbb{R}$ be continuous. If $f(a) < 0 < f(b)$ or $f(a) > 0 > f(b)$, then there exists a $c \in (a, b)$ where $f(c) = 0$.

The Intermediate Zero Theorem is the specific case of the Intermediate Value Theorem when $L = 0$. This is the version relevant to the process of locating roots of functions. We use the abbreviation IVT to refer to either statement, although it is the second one we attend to more closely.

7.3.1 Differentiating Conditions in Statements

As a general rule, mathematicians try to avoid ambiguity and inefficiency. Everything required should be explicitly stated, and everything explicitly stated should be required. In terms of style, mathematicians lean toward brevity, saying what is necessary in the fewest words needed. Because mathematical concepts build on each other, phrases and concepts can include hidden implications. Unpacking the full meaning of a mathematical statement involves reviewing the relevant definitions and their implications as well as paying attention to the statement's logical structure.

Turning our attention to the IVT, the first thing to point out is the inclusion of continuity as a condition. Without it, the conclusion does not hold (see Fig. 7.2). We should also acknowledge the phrase "there exists" in the theorem's conclusion. The existence of a value c does not preclude the possibility that there could be more—existence and uniqueness are different questions. These two observations

Fig. 7.2 A discontinuous function f, defined on $[1, 9]$, with $f(1) < 0 < f(9)$ but no zeros in $(1, 9)$

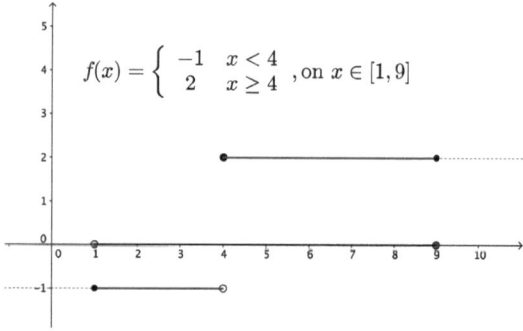

about the IVT were raised in our earlier discussion of the teaching scenario. What other aspects of the IVT might be interrogated?

A useful heuristic to better understand the conditions of a theorem is to ask what happens if the conditions were changed or if parts were left out.[2] Consider the assumption, $f(a) < 0 < f(b)$. As stated, it utilizes strict inequalities. What happens if we change it to $f(a) \le 0 \le f(b)$? You might consider some possibilities before moving on. One example to test would be when $f(a) = 0$ and $f(b) > 0$. In this case, a line between $(a, 0)$ and $(b, f(b))$ does not have any zeros in (a, b) and so the conclusion would not hold. What if both endpoints were zero? In this situation we might draw a sine curve that crosses the x-axis several times in the interval, or a parabola with roots at a and b that has no zeros in the interval. This latter example reinforces the prior observation that the IVT's conclusion no longer follows with the amended conditions. To fix this we could edit the conclusion to assert the existence of a value $c \in [a, b]$ instead of $c \in (a, b)$. This puts us on firm logical ground, but the cases when either $f(a)$ or $f(b)$ equal zero make the conclusion of the IVT rather trivial. The takeaway of this experiment is that the use of either strict or inclusive inequalities in the condition and the conclusion are no coincidence—they are linked. And the strict inequalities yield the most appropriate version of the IVT!

Let's look more carefully at one other condition: $f : [a, b] \to \mathbb{R}$. There is a component to this part of the hypothesis that is often overlooked; in particular, it says the domain is a closed interval. What happens if we remove this part of the condition, but keep everything else the same? Does the conclusion about the existence of a zero still hold?

Question *Let $f(x) : A \to \mathbb{R}$ be continuous on its domain A, and let a and b be points in the domain with $a < b$. If $f(a) < 0 < f(b)$ or $f(a) > 0 > f(b)$, is it true f must have at least one zero in (a, b)?*

---

[2] Brown and Walter [3] describe this as the "what-if-not" strategy for problem posing.

Fig. 7.3 A continuous function $f : [1, 4) \cup (4, 9] \to \mathbb{R}$, with $f(1) < 0 < f(9)$, but no zeros in $(1, 9)$

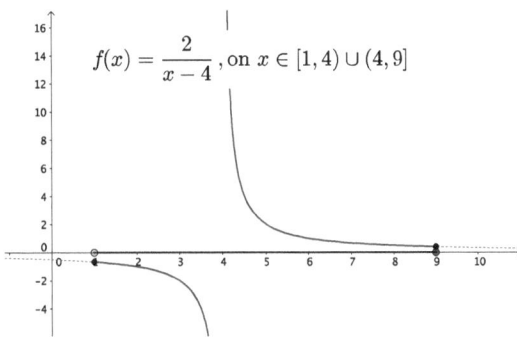

$$f(x) = \frac{2}{x-4}, \text{ on } x \in [1, 4) \cup (4, 9]$$

To answer the question, let's think about whether it is possible to construct a continuous function with, for example $f(a) < 0 < f(b)$, but that has no zeros in the interval (a, b). This requires us to stretch our image of a function. (The examples in Fig. 7.1 might be a guide.) We also have to sharpen our understanding of what it means to assert that a function is continuous over a given set A. In Chap. 6 we learned that, according to the standard definition of continuity, a function such as $g(x) = \frac{1}{x}$ is continuous over its natural domain $A = \{x \in \mathbb{R} : x \neq 0\}$. Likewise, the related example $f(x) = \frac{2}{x-4}$ depicted in Fig. 7.3 is continuous on the domain $A = [1, 4) \cup (4, 9]$. Note that f satisfies $f(1) < 0 < f(9)$ but has no zeros in $(1, 9)$. This example shines a spotlight on a condition in the IVT that is subtly embedded in the notation: the domain of the function must be a closed interval $[a, b]$. In our example, f is continuous at every point in A and thus continuous on A, but f is not defined at $x = 4$. Being defined at a point is a prerequisite for continuity at that point; we need to distinguish between the two conditions and interrogate them separately.

7.3.2 Use of Conditions in Proofs

Mathematical propositions are typically crafted so that all the conditions in the hypothesis are required for the conclusion to follow. Granted, there are certainly exceptions. Teachers and textbook authors sometimes include additional information to make the statements more understandable to students; sometimes additional conditions are added to simplify the proof. Generally-speaking, however, we should presume that every condition is required for the proof to go forward—that the conditions do not include unnecessary information.

Let's consider the two particular conditions of the IVT we delineated in the previous discussion: (i) f is *defined* on $[a, b]$; and (ii) f is *continuous* on $[a, b]$. Below is a standard proof of the IVT that uses the Nested Interval Property (see Abbott [1] on pp. 138–139). As you read the proof, identify where each assumption comes into play. At what point does the argument break down if f is not defined on $[a, b]$? Where does it break down if f is not continuous on $[a, b]$?

Fig. 7.4 With f not defined
on $[a, b]$, the constructed
sequence of nested intervals
may stop

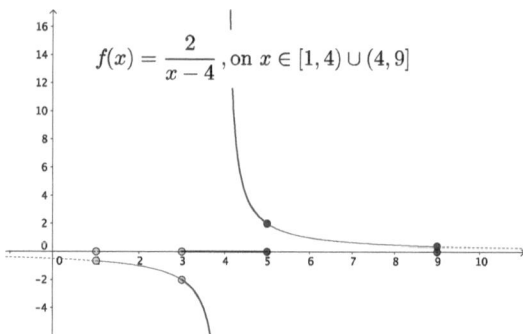

$$f(x) = \frac{2}{x-4} \text{, on } x \in [1, 4) \cup (4, 9]$$

Proof We will consider the case when $f(a) < 0 < f(b)$, and begin with the
interval $I_0 = [a, b]$. At its midpoint, z, test the value of the function. If $f(z) \geq 0$,
set $a_1 = a$ and $b_1 = z$; if $f(z) < 0$, set $a_1 = z$ and $b_1 = b$. This process can
be repeated on the new interval $I_1 = [a_1, b_1]$ to create $I_2 = [a_2, b_2]$. Continuing
this procedure, we construct a sequence of nested intervals $I_n = [a_n, b_n]$ with the
property that $f(a_n) < 0$ and $f(b_n) \geq 0$ for every n. Because we are bisecting each
time, the length of each interval is half the length of the preceding one, which means
the lengths are converging to 0.

By the Nested Interval Property (Abbott's Theorem 1.4.1), the intervals all
contain at least one point c. Because c belongs to every interval and the lengths
of the intervals tend to 0, the sequences of left- and right-hand endpoints both
approach c, meaning $(a_n) \rightarrow c$ and $(b_n) \rightarrow c$. Therefore, $f(a_n) \rightarrow f(c)$ and
$f(b_n) \rightarrow f(c)$. By construction, we know that $f(a_n) < 0$ for every n and so
$f(c) = \lim f(a_n) \leq 0$ (see Theorem 2.3.4 in Abbott). Likewise, $f(b_n) \geq 0$ for
every n and so $f(c) = \lim f(b_n) \geq 0$. Since $f(c) \leq 0$ *and* $f(c) \geq 0$, we know
$f(c) = 0$.   □

Let's focus first on the condition that f is defined at every real number in $[a, b]$.
Where does the argument fail without this assumption? We want to think specifically
about the values in the domain, and what it means for a function to be, or not
be, defined at a particular value. The argument entails constructing a sequence of
intervals $I_0, I_1, I_2, \ldots$ where at each stage we evaluate the midpoint z to determine
the endpoints of the next interval. The problem occurs if z is not in the domain—if
we can't compute $f(z)$ then there is no way to generate the next interval. Figure 7.4
gives an example where this occurs. The first midpoint, $z = 5$, makes $I_1 = [1, 5]$;
the next midpoint, $z = 3$, makes $I_2 = [3, 5]$; but the function is not defined at the
next midpoint, $z = 4$. Hitting this roadblock, the process stops. We are not able to
generate the sequence of intervals I_n which are required to produce the value of c
satisfying the conclusion of the theorem.

Where in this argument does the assumption of continuity on $[a, b]$ come into
play? As long as f is defined at every real number in $[a, b]$, the bisection algorithm
will result in an infinite sequence of intervals I_n. Nothing about this aspect of

Fig. 7.5 With f defined but not continuous on $[a, b]$, the sequence of function values of the left-endpoints $f(a_n)$ does not necessarily approach $f(c)$

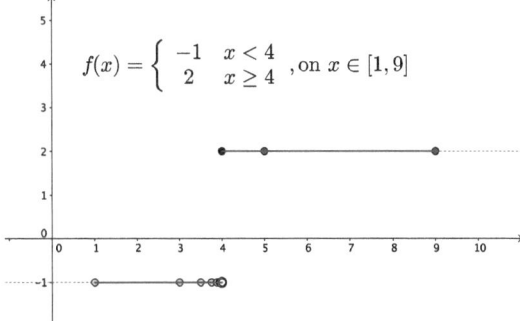

$$f(x) = \begin{cases} -1 & x < 4 \\ 2 & x \geq 4 \end{cases}, \text{on } x \in [1, 9]$$

the proof requires continuity. It is only toward the end of the proof, when we are considering the sequences of left-endpoints and right-endpoints, that continuity is needed. With a left-endpoint sequence $(a_n) \rightarrow c$, it is the continuity of f that allows us to conclude $f(a_n) \rightarrow f(c)$. The same is true of the right-endpoint sequence (b_n). The example in Fig. 7.5 shows how this process breaks down without continuity. It illustrates that for a function with a jump discontinuity, the algorithm in the proof successfully generates a sequence of nested intervals with a unique point of intersection. In this example, the left-hand endpoint sequence (a_n) converges to $c = 4$ as does the right-hand endpoint sequence (b_n). Without continuity, however, we are no longer guaranteed that $f(a_n)$ and $f(b_n)$ both converge to $f(c)$. For this example, $f(a_n) \rightarrow -1$ while $f(b_n) \rightarrow 2$; the limits are different and, notably, not equal to 0.

7.4 Connecting to Secondary Teaching

In the initial teaching situation, the student provided a reasonable summary of the IVT, but one that made some implicit assumptions about functions. In a classroom context, we should not necessarily regard students' implicit assumptions to be wrong or unwarranted. Rather, we see them as an inherent part of the learning process. It is the teacher's responsibility to unearth these assumptions and call attention to them (TP.1). In the scenario, the student's statement is true for polynomials because these functions are *defined* and *continuous* on $\mathbb{R}$—polynomials meet the conditions of the theorem. Details like these allow teachers to illuminate the nuances of mathematical relationships and clarify why certain statements are valid. Doing this in an effective way requires expanding the kinds of examples available to students so that they understand the need for being explicit about the relevant details.

7.4.1   Implicit Assumptions in the Classroom

The Intermediate Value Theorem is frequently included as a topic of secondary school mathematics. When working with the IVT, the function needs to be both defined and continuous on the interval $[a, b]$, and we want to ensure that students understand both of these conditions. There cannot be "holes" in the domain, nor "jumps" in the function.

Through no fault of their own, secondary mathematics students are especially susceptible to making implicit assumptions about functions. At the university level, mathematics students encounter discontinuous functions frequently enough that they become an organic part of their example space. From this perspective, the hypothesis of continuity in the IVT stands out as an explicit and necessary requirement. From a secondary student's perspective, it may feel as though there is no need to specify that the function be continuous because all the functions they work with are continuous. Adding the modifier "continuous" becomes optional if every function is assumed to have this property already. To take another example, if the term "pyramid" calls to mind only solids with square bases, why would a student think there is a need to specify "square pyramid"? As teachers, we should consider not only whether a student's statement is valid in general, but whether it would be valid under their implicit assumptions. If so, a good response should make clear that the student's statement is relatively correct, and then reveal the assumptions that the student left unsaid (TP.1). Doing so might involve the teacher introducing an example that does not conform to the missing assumptions, an instance of TP.2.

To make this concrete, let's return to the original teaching scenario and consider the following response:

> The next time the class meets, Mr. Cai brings Chrissy's exit ticket to their attention. After discussing it, Mr. Cai concludes:
>
>> Chrissy's description is a great way to generalize how the IVT works *for polynomials*. This is true because polynomials are both *defined* and *continuous* on the real numbers. Although we will primarily use the IVT with respect to polynomials, it is important to note that the conclusions would not hold in all situations.
>
> Mr. Cai then asks the class to come up with and graph examples of functions that take on both positive and negative values but are not continuous, or not defined, everywhere. Using their examples, he asks them to check whether those examples always have a root between two values with opposite signs.

Teaching typically involves taking concepts apart in order to make the various components clear, and then organizing those components into a useful order for

Fig. 7.6 An IVT-based argument for the existence of acute, obtuse, and right isosceles triangles

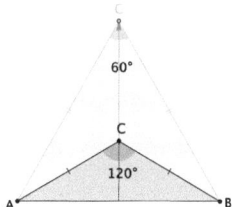

learning.[3] This task of unpacking can be more difficult than it sounds. Not only do teachers typically compress their own knowledge as they progress through more advanced topics, but by its nature mathematics is a condensed discipline. Each word and symbol in a statement like $f : [a, b] \rightarrow \mathbb{R}$ may convey some detail that needs to be fleshed out. As the notation and concepts become more familiar to us, we can lose track of the different layers and forget what it feels like to be encountering the ideas for the first time. Being attentive to this starts with paying attention to *all* of the mathematical ideas expressed in a statement and anticipating where students might reasonably make assumptions based on what they know. Teachers can then make decisions about what needs elaboration before moving forward to order, structure, and connect the individual pieces into a meaningful and coherent whole. Students' notions of a concept can be refined, and further developed, only once their sense of the concept has been expanded first.

7.4.2 Implicitly Assuming the IVT in Secondary Mathematics

To this point we have focused on the implicit assumptions students might make about the hypothesis of the IVT, but it is not unusual in secondary school mathematics classrooms for students to implicitly assume the *entirety* of the theorem. The content of the IVT is so plausible that on occasion it can be unwittingly invoked in the course of some other argument. Here is an example from a geometry class, where the students were asked whether isosceles triangles can also be acute, right, and obtuse triangles. A student in the class responds with the following argument based on the drawing in Fig. 7.6:

> I am thinking about the vertex of an isosceles triangle. If it is down low, the triangle would have an obtuse angle, and if it is dragged further up along this center line, the angle would be acute. So, somewhere in between, it must be exactly 90°. Like, if the bottom angle is 120°, and the top 60°, then 90° is right in the middle so it's probably exactly halfway between those two. So, yes, there are obtuse, acute, and right isosceles triangles.

---

[3] Ball and Bass [2] describe this notion of unpacking; teachers need to be able to "deconstruct [their] own mathematical knowledge into less polished and final form, where elemental components are accessible and visible" (p. 98).

Fig. 7.7 Continuity of the
function $f(h)$ which assigns
an angle measure to each
height

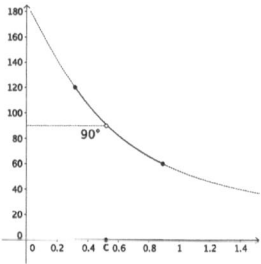

Not everything in the student's argument is correct, but the conclusion that a right isosceles triangle exists somewhere in between the two examples depicted is indeed true. The student's reasoning implicitly draws on the IVT. Let's think more about why this is the case.

First, the student has observed an angle value of 120° at one location and an angle value of 60° at another. This is akin to observing values $f(a) = 120$ at location a, and $f(b) = 60$ at location b, where f is measuring the size of the angle. The student then argues that, since 90° is between 120° and 60°, there must be a triangle where the 90° angle measure is attained. The student makes no mention of the IVT, or even of a function f, but there is an unmistakable impression that the spirit of the IVT is being invoked. Filling in the details to put this argument on solid ground requires thinking a little more about the angle-measuring function f.

To define a proper domain for f, we focus on the perpendicular bisector from the diagram in Fig. 7.6. This is the vertical line along which the student was mentally "dragging" the central vertex, and we define the *height h* to be the distance from the horizontal base to this imagined vertex. The values of h are the input values of our function $f(h)$, which we formally define to be the measure of the central angle of the isosceles triangle with vertices A, B and height h.

If the base between A and B has length 1, then height $a \approx 0.28$ yields $f(a) = 120$ and height $b \approx 0.87$ yields and $f(b) = 60$, which are the two depicted in the figure from above. (The exact values are $a = \frac{1}{2\sqrt{3}}$ and $b = \frac{\sqrt{3}}{2}$). This provides the necessary raw material to properly apply the IVT. (See Fig. 7.7.) The function f is *defined* for all values of h in the closed interval $[a, b]$. To convince ourselves that f is *continuous*, we mentally drag the vertex up and down the perpendicular bisector, just as the student did, and observe there are no jumps or holes. (For a more rigorous argument, we can deduce $f(h) = 2 \arctan\left(\frac{1}{2h}\right)$ and appeal to the continuity of the inverse trigonometric functions.) Since $120 > 90 > 60$, a straightforward application of the IVT confirms the existence of a height c where $f(c) = 90$.

What the IVT does not tell us is how to compute the value of c. On this point, the student's hunch that c is "halfway between" $a \approx 0.28$ and $b \approx 0.87$ is off the mark. The student seems to have made the additional implicit assumption that $f(h)$ is linear, which we can see from Fig. 7.7 is not the case. For what it's worth, $f(c) = 90$ when $c = 0.5$, which is in the interval $[0.28, 0.87]$ but not at its midpoint.

Problems

7.1 Graph the function $f(x)$ shown below, using any preferred technological tool:

$$f(x) = \begin{cases} 2x & x < 0 \\ \frac{1}{8}x(x-1)^2 + x(x-2)(x-5) & 0 \le x < 5 \\ 10 & 5 \le x \end{cases}$$

First, show that $f(x)$ meets the conditions of the IVT on the interval $[-1, 5]$. Second, what range of values can you be certain that $f(x)$ takes on due to the IVT? Third, does $f(x)$ take on any additional values on the interval $[-1, 5]$? If so, what are they? How do you know? Fourth, how many zeros does the IVT guarantee $f(x)$ has on $[-1, 5]$? How many zeros does $f(x)$ actually have on this interval? Which zero(s) would the nested intervals process in the proof of the IVT find (beginning with $I_0 = [-1, 5]$)?

7.2 The following multiple choice question was on a geometry test:

Quadrilateral $ABCD$ is a rectangle, with diagonal AC. How do the quantities $\frac{AC}{AB}$ and $\frac{AB}{AD}$ compare? a) $\frac{AC}{AB} > \frac{AB}{AD}$; b) $\frac{AC}{AB} < \frac{AB}{AD}$; c) $\frac{AC}{AB} = \frac{AB}{AD}$; d) The relationship cannot be determined from the given information. Justify your answer.

First, determine your answer to the question. Next, consider a student's written response: "The answer is (a) because $\frac{AB}{AD} < 1 < \frac{AC}{AB}$." Discuss any assumptions the student may be making about the situation. Under those assumptions is the student's statement valid? Last, describe how you would respond to the student as a teacher, making explicit under what assumptions the student's statement is correct and providing examples that do not conform to those assumptions.

7.3 The following question was on a geometry test: "A triangle has side lengths of 3 and 4 units. What do you know about the third side length?" One student drew a picture of a 3-4-5 right triangle, and wrote: "I know that $a^2 + b^2 = c^2$. And $3^2 + 4^2$ is equal to 5^2. The third side must be 5." Discuss any assumptions that the student may be making about the situation. Under those assumptions is the statement valid? Then, describe for what concept the student's "concept image" appears to be limited. Discuss how you might push the student to expand their sense of that concept so that they would recognize the limitation that arose from their implicit assumption.

7.4 A geometry class is learning about the surface area and volume of geometric solids. The teacher provides the following formulas for the volume and surface area of a regular polygonal pyramid—a pyramid with a regular n-gon as its base:

- $V_{reg\ poly\ pyr} = \frac{1}{3}B^*h$, where B^* is the area of the n-gon's base, and h the height of the pyramid
- $SA_{reg\ poly\ pyr} = B^* + \frac{1}{2}P^*l$, where B^* is the area of the n-gon's base, P^* the perimeter of the n-gon, and l the slant height of the triangular face (the slant height is the height of a triangular face of a pyramid—the length of the line segment along a triangular face from the base to the apex of the pyramid).

The formulas given by the teacher for the volume and surface area of a pyramid have some specific assumptions. What assumptions about pyramids are explicitly included in the description of volume? In the description of surface area? Do either of the descriptions about volume or surface area have any implicit assumptions about pyramids—things unstated but assumed about pyramids for these formulas to be valid? Discuss any implicit assumptions and how they would have an impact on the pertinent formulas for volume or surface area.

7.5 A student is attempting to split a triangle XYZ into two equal-area pieces:

Well, I'm not sure if it is *exactly* in two equal pieces as I have drawn it. But I could adjust the point (C) somewhere along that segment (YZ) and they would be.

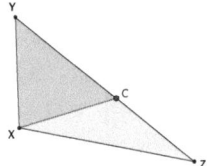

How has the following student's statement implicitly used the Intermediate Value Theorem? Verify all necessary conditions of the IVT are met—it is okay to be somewhat informal for continuity in this case. Discuss which mathematical ideas you might make explicit to students, and how you might do so. [This exercise was adapted from a classroom example in [4].]

7.6 For hourly workers, work after 40 h typically results in an overtime rate of 1.5 times the normal hourly rate. Suppose a student observes in this situation that at 30 h, a worker would be earning an hourly rate of, say, $20, and at 50 h they would be earning an hourly rate of $30. The student suggests that the worker must have been earning an hourly rate of exactly $25 at some point. First, describe how the student's reasoning has implicitly assumed the IVT. Second, the student's conclusion is false. Describe what conditions of the IVT have not been met.

7.7 In this chapter it was useful for us to think about isolating conditions. Indeed, identifying whether there were functions with a certain set of properties was productive (e.g., a function $f(x) : A \to \mathbb{R}$ that is continuous on its domain A, with $f(a) < 0 < f(b)$, but has no zero in the interval (a, b)). In Exercise 4.4.8, Abbott asks a similar kind of question: "Give an example of each of the following, or provide a short argument for why the request is impossible. (a) A continuous function defined on $[0, 1]$ with range $(0, 1)$..." Questions such as these exemplify what teaching principle? Explain your reasoning.

7.8 The Intermediate Value Theorem (IVT) is a statement about continuous functions; namely, that if a function f is continuous on $[a, b]$, then on that interval the function will attain all values between $f(a)$ and $f(b)$. Abbott also defines what he calls the Intermediate Value Property (IVP), in Definition 4.5.3 [1, p. 139].

> A function f has the *intermediate value property* on an interval $[a, b]$ if for all $x < y$ in $[a, b]$ and all L between $f(x)$ and $f(y)$, it is always possible to find a point $c \in (x, y)$ where $f(c) = L$.

Think about these two statements in relation to TP.3, about exposing logic. Rephrase the IVT to incorporate the IVP, and discuss any pertinent observations about the various properties of functions in terms of logic.

Turning the Tables

Reflecting on *teaching* from your *learning* in real analysis: TP.6

To reflect more on TP.6—seeking out and giving multiple explanations—we include some additional commentary about the description and proof of the Intermediate Value Theorem, as given in Abbott's text.

The teaching principle advocates using multiple explanations for the same phenomenon because some students may follow one explanation better than another. As an example of this practice, after introducing the content of the IVT, Abbott goes on to state a topological theorem about the preservation of connected sets: "Let $f : G \to \mathbb{R}$ be continuous. If $E \subseteq G$ is connected, then $f(E)$ is connected as well" (Theorem 4.5.2). He then explains the IVT is really just a special case of this theorem. In other words, there is a typical analysis approach that accounts for the phenomenon in the IVT, but there is also a topological approach that is more general. Here, we see TP.6 exemplified in terms of providing two descriptions of an observed phenomenon. Describing ideas in multiple ways—and from multiple mathematical domains—adds depth to the mathematics being studied.

We see this teaching principle again in Abbott's justification of the IVT when he provides two different proofs (pp. 138–139). The one given in this chapter is based on the Nested Interval Property, but Abbott gives another that uses the Axiom of Completeness. For this second proof, Abbott defines a set $A = \{x \in [a, b] | f(x) \leq 0\}$. Because A is bounded, the Axiom of Completeness asserts it must have a least upper bound, c. He then shows $f(c)$ must be equal to 0. The two proofs provide two different arguments for readers to follow, creating opportunities for them to make connections across the two approaches. Together, they provide a more comprehensive sense of the IVT and why it is true. In the spirit of TP.6, Abbott's text utilizes the pedagogical approach of describing a phenomenon in multiple ways and justifying it in multiple ways as well. Because of this approach, we have a richer sense not only of what the IVT means, but why it is true.

References

1. Abbott, S. (2015). *Understanding analysis* (2nd ed.). New York, NY: Springer.
2. Ball, D. L., & Bass, H. (2000). Interweaving content and pedagogy in teaching and learning to teach: Knowing and using mathematics. In J. Boaler (Ed.), *Multiple perspectives on mathematics teaching and learning* (pp. 83–104). Westport, CT: Ablex.
3. Brown, S. I., & Walter, M. I. (2005). *The art of problem posing* (3rd ed.). Mahwah, NJ: Lawrence Erlbaum.
4. Jakobsen, A., Thames, M. H., Ribeiro, C. M., & Delaney, S. (2012). Using practice to define and distinguish horizon content knowledge. In *Proceedings of the 12th International Congress on Mathematical Education* (pp. 4635–4644). Seoul, South Korea: International Commission on Mathematical Instruction.
5. Tall, D., & Vinner, S. (1981). Concept image and concept definition in mathematics with particular reference to limits and continuity. *Educational Studies in Mathematics*, *12*(2), 151–169.

Continuity, Strict Monotonicity, Inverse Functions and Solving Equations

8

8.1 Statement of the Teaching Problem

There are two critical ideas that students learn when first solving algebraic equations. The first is "do the *same* thing, to *both* sides." If two sides of a scale are currently balanced, adding or removing something on one side should be counteracted by doing the same thing on the other side to maintain equilibrium (check out "algebra balance scales," a common manipulative). The second idea is that the algebraic process for solving an equation amounts to systematically "undoing" operations while maintaining the relation of equality.

The undoing process of solving equations is linked to *inverses*, or performing inverse operations. To undo a product, for example, we divide both sides of an equation by the given factor (or multiply by the reciprocal of that factor) because multiplication and division are inverse operations. In more general terms, this undoing process is connected to inverse functions, which are functional inverses (not additive or multiplicative ones). The use of inverse functions to undo aspects of equations becomes more apparent as the equations get more complex.

Consider the following pedagogical situation:

> Ms. Gonzalez is discussing how to solve trigonometric equations with her class. A student, Meng, goes to the board, and explains their solution to $\sin(2x) = 0.7$:
>
> First, I took the 'arcsin' of both sides, resulting in:
>
> $$\arcsin(\sin(2x)) = \arcsin(0.7)$$

(continued)

N. H. Wasserman et al., *Understanding Analysis and its Connections to Secondary Mathematics Teaching*, Springer Texts in Education,
https://doi.org/10.1007/978-3-030-89198-5_8

Then, since 'sin' and 'arcsin' cancel out, $2x$ remains on the left, and I computed arcsin (0.7) on my calculator and got around 0.7754. So we have:

$$2x = 0.7754$$

Then, I divided by 2 to get $x = 0.3877$, and then added '$+2\pi k$', so that we have:

$$x = 0.3877 + 2\pi k$$

In this example, the student appears to follow the heuristic of doing the same thing to both sides, "undoing" operations until the variable x is isolated. The student even appears to recognize that trigonometric equations are a little different by their inclusion of '$+2\pi k$' at the end. Both of these are positive reflections, but it is important to consider whether *all* the solutions to the initial equation have been identified. Is there anything missing in the student's process? Are there any limitations that might need to be addressed (TP.1), or places where their logic might be incomplete (TP.3)? Analyzing a student's mathematical work is a particular challenge of teaching. As it turns out, in the context of solving equations, inverse functions have some nuances that require further attention.

Before moving on, try to solve the trigonometric problem for yourself. Then think about how you would respond to the student, or the class, if you were the teacher.

8.2   Connecting to Secondary Mathematics

8.2.1   Problematizing Teaching and the Pedagogical Situation

To unpack this scenario, let's write out the student's answer as the following list of approximate solutions:[1]

$$x \approx \ldots, -12.1787, -5.8955, 0.3877, 6.6709, 12.9541, \ldots$$

Now, consider $x \approx 1.1831$, which is *not* on this list. Based on the original equation, we can compute: $\sin(2 \cdot 1.1831) = \sin(2.3662) \approx 0.7$. That is, 1.1831 appears to be another solution to the equation, yet one not identified by the student. The goal is to find all solutions, but some look to be missing. What went wrong? Try to identify the two mistakes in the student's work.

Prior to assigning a problem like this, teachers probably know what the solutions are, either because they worked them out themselves or found them in the teacher's

---

[1] Throughout the chapter, we use decimal approximations like the student did. Actual values (e.g., arcsin $(0.7) + 2\pi$) are cumbersome and may obscure the points being made.

manual. A first reaction might be to acknowledge what the student has done well. Although adding '$+2\pi k$' violates the "do the same thing to both sides" rule, it does so in a good way, and in all other aspects the student has attended to this principle. At the same time, the student is missing some solutions. Perhaps a second reaction would be to tell the student of their mistake and indicate, like we did, one of the missing solutions. Or, a teacher might tell the student something more direct, like, "Add 'πk,' not '$2\pi k$'." But doing so would be giving a rule without any accompanying mathematical explanation, going against TP.5.

8.2.2 Solving Trigonometric Equations

Ideas related to symmetry and periodicity are at the core of the student's errors and are critical to consider when using inverse functions to solve trigonometric functions.

Inverse functions are an amalgam of two ideas: "inverse" and "function." As described in Chap. 7, the formal definition for a function $f : A \rightarrow \mathbb{R}$ is a set of points (x, y) in the plane where each x in the domain A appears as the first entry in exactly one ordered pair. In the more familiar function notation, each ordered pair (x, y) is written $f(x) = y$ to indicate that f maps the domain point x to the range point y. The *inverse function* is the mapping that goes in the reverse direction—the inverse should map each range point y back to the original x. In the ordered pair notation, this means that if (x, y) is a point on f then (y, x) should be a point on its' inverse.

A problem arises if the original function f maps two distinct values in its domain to the same point in the range. The function $f(x) = x^2$ has this property. Both $(2, 4)$ and $(-2, 4)$ are among its ordered pairs because both $2^2 = 4$ and $(-2)^2 = 4$. But when we consider the inverse of f, we are left with a problem because $(4, 2)$ and $(4, -2)$ cannot both be included in the inverse without violating the definition of function. The domain value 4 requires a unique output and so can only be mapped to one range value. The crux of the matter is *not every function has an inverse function*!

The solution to this problem of finding an inverse function when none exists is to modify the original function by restricting its domain so that it *is* possible to obtain an inverse function. Although $f(x) = x^2$ does not have an inverse when we consider it as a function on $\mathbb{R}$, it does have an inverse if we shrink the domain to $[0, \infty)$. In fact, the inverse is the familiar function $g(y) = \sqrt{y}$ whose range is, by convention, $[0, \infty)$. Note that $g(f(x)) = x$, which means we can use g to "undo" f, but *only* for values of x in the restricted domain $[0, \infty)$. If $x = -2$, for instance, $g(f(-2)) \neq -2$. Like $f(x) = x^2$, the trigonometric function $\sin(x)$ does not have an inverse when considered as a function on all of $\mathbb{R}$. The convention in the case of the sine function is to restrict the domain to the interval $[-\pi/2, \pi/2]$. On this smaller domain, it is possible to define an inverse function, which is typically denoted as $\arcsin(y)$. The function $\arcsin(y)$ "undoes" the sine function, meaning $\arcsin(y)$ outputs the angle x whose sine is y. But just as with the square root function, we

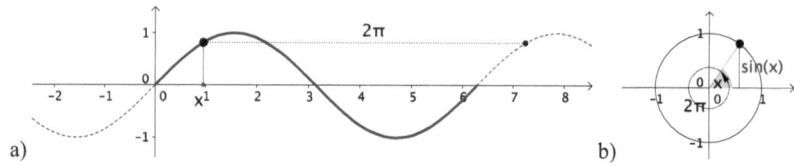

Fig. 8.1 Periodicity of sin (x) from its (**a**) graph, and (**b**) unit circle

have to be careful about domain concerns. The relationship $\arcsin(\sin(x)) = x$ holds, but only for values of x in the agreed upon domain $[-\pi/2, \pi/2]$. In the classroom scenario, Meng's assertion that $\arcsin(\sin(2x)) = 2x$ is therefore only valid if $2x \in [-\pi/2, \pi/2]$, or $x \in [-\pi/4, \pi/4]$. The student finds a correct solution, but they miss any that fall outside of this restricted interval. Can we recover the missing solutions?

In the case of trigonometric functions we can. One reason is because trigonometric functions are *periodic*. The outputs repeat at a fixed interval known as the period; the period of sin (x) is 2π. This means for any x, the sine value will be the same at $x + 2\pi$, $x + 4\pi$, etc., because these angles represent the same point on the unit circle, where the sine is just the y-coordinate. This periodicity is apparent in Fig. 8.1.

In terms of the student's work from the pedagogical situation, the moment Meng takes the arcsine of both sides, the resultant '$2x \approx 0.7754$' refers to an angle that we now know lies in $[-\pi/2, \pi/2]$. One way to improve on Meng's solution would be to invoke periodicity at this earlier point in the computation and try to recover the missing solutions by writing '$2x \approx 0.7754 + 2\pi k$'. Dividing both sides by 2 gives '$x \approx 0.3877 + \pi k$'. The $2\pi k$ Meng added at the end missed half of the solutions! Another approach would be to recognize that the function $g(x) = \sin(2x)$, from the original equation, has a period of π (not 2π)—the '2' increases the frequency of the oscillations, cutting the period in half. Thus, we could add '$+\pi k$' after arriving at '$x \approx 0.3877$' to account for the periodicity of the function in the original equation.[2]

Even with this modification, it turns out there are still solutions missing from our list. In addition to periodicity, trigonometric functions also possess other *symmetrical* aspects. Similar to the way -2 and 2 both have a square of 4, for any y-value strictly between -1 and 1 there are *two* values of x within a single period of the sine satisfying $\sin(x) = y$. This symmetry can be seen in both representations in Fig. 8.2.

In particular, for any x in $(-\pi/2, \pi/2)$, the angle $\pi - x$ is another angle within the same period that has the same sine value (i.e., $\sin(x) = \sin(\pi - x)$). Just as $\sqrt{4}$ produces only 2 (and not also -2), $\arcsin(0.7)$ yields *one* angle, 0.7754, with a sine

---

[2] To formalize this, $g(x) = \sin(2x)$, on $[-\pi/4, \pi/4]$, has an inverse g^{-1} (which is *not* $\arcsin(y)$). Because g has a period of π, the step $g^{-1}(g(x)) = g^{-1}(0.7)$ simplifies to $x = g^{-1}(0.7) + \pi k$. What is g^{-1} though? It turns out the inverse function of g can be expressed as $g^{-1}(y) = 0.5\arcsin(y)$.

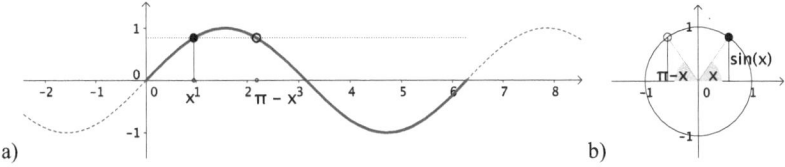

Fig. 8.2 Symmetry of sin (x) from its (**a**) graph, and (**b**) unit circle

value of 0.7 but omits another—namely, $\pi - 0.7754 \approx 2.3662$. Because this new solution is within the same period as the original 0.7754 solution, it does not get picked up when we account for periodicity by considering '$2x \approx 0.7754 + 2\pi k$.' Along with '$2x \approx 0.7754 + 2\pi k$,' we need to include '$2x \approx 2.3662 + 2\pi k$' as another set of solutions. This means '$x \approx 1.1831 + \pi k$' are also solutions to the equation.

8.2.3 Inverse Functions

For a given function f, we use f^{-1} to denote its inverse function.[3] What properties must f possess to guarantee that it has an inverse function f^{-1}?

Here is a list of some likely familiar properties to consider:

- A function $f : A \rightarrow \mathbb{R}$ is **one-to-one** (*injective*) if taking distinct elements $x_1 \neq x_2$ in the domain A implies $f(x_1) \neq f(x_2)$ in $\mathbb{R}$.
- A function $f : A \rightarrow B \subseteq \mathbb{R}$ is **onto** B (*surjective*) if $f(A) = B$, meaning for any element $y \in B$ there is an $x \in A$ for which $f(x) = y$.
- A function $f : A \rightarrow \mathbb{R}$ is **strictly increasing** on A if $f(x_1) < f(x_2)$ whenever $x_1 < x_2$ (and $x_1, x_2 \in A$), and **strictly decreasing** on A if $f(x_1) > f(x_2)$ whenever $x_1 < x_2$ (and $x_1, x_2 \in A$). A **strictly monotonic** function is one that is either strictly increasing or strictly decreasing on its domain.[4]

In general, a function must be *one-to-one* to have an inverse (cf. Abbott's [1] exercise 4.5.8). As long as distinct elements of the domain are mapped to distinct elements in the range, there is no ambiguity about how to define the inverse on the range values. A function that is *not* one-to-one, however, would have two elements, $x_1 \neq x_2$, where $f(x_1) = f(x_2) = y$ for some y. In this case, $f^{-1}(y)$ would somehow need to output both x_1 and x_2, which precludes f^{-1} from being a function.

---

[3] This is standard notation despite the potential ambiguity that f^{-1} might suggest the multiplicative inverse $1/f$ rather than the functional inverse.

[4] We have chosen to define *strictly monotonic*, and not *monotonic*. A monotonic function, instead of strict inequalities, would have $f(x_1) \leq f(x_2)$ and $f(x_1) \geq f(x_2)$. As a result, a horizontal line is monotonic. "Non-decreasing" and "non-increasing" are the most apt descriptors—a monotonic function is one that is either non-decreasing or non-increasing.

Fig. 8.3 A non-strictly
monotonic function that has
an inverse function

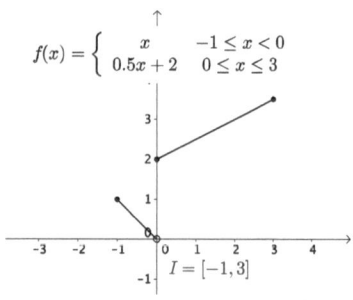

$$f(x) = \begin{cases} x & -1 \le x < 0 \\ 0.5x + 2 & 0 \le x \le 3 \end{cases}$$

$I = [-1, 3]$

In the event that f is *not* one-to-one, the standard resolution is to restrict the domain
to a smaller set on which f *is* one-to-one.

Definition Given a one-to-one function f with domain A and range $B = f(A)$,
the **inverse function** f^{-1} is the function with domain B and range A given by
$f^{-1} = \{(y, x) | (x, y) \in f\}$; that is, $f^{-1}(y) = x$ where $y = f(x)$.

Two immediate consequence of this definition are the cancellation equations used
in the previous example: $f^{-1}(f(x)) = x$ for all $x \in A$ and $f(f^{-1}(y)) = y$ for all
$y \in B = f(A)$.

In the context of secondary mathematics, the majority of functions are continuous
and have a domain of $\mathbb{R}$ or some interval $I \subseteq \mathbb{R}$. In this situation, the existence
of an inverse can be considered more specifically in terms of continuity and strict
monotonicity. As we have already seen, assuming f is continuous on an interval
does not guarantee an inverse function; recall the example $f(x) = x^2$ on, say,
$[-3, 3]$. The crucial ingredient needed to produce an inverse is strict monotonicity.
As we will prove, a function that is defined and strictly monotonic on an interval I
will always have an inverse function.

The converse is not true; functions that have inverse functions are not always
strictly monotonic. That is to say, a function that is not strictly monotonic may
still have an inverse function. (Can you think of an example?) Figure 8.3 shows a
non-strictly monotonic function that has an inverse function. Notably, this boundary
example possesses a discontinuity. This is unavoidable. If a function f is continuous
at every point of an interval I (as is the case with the most functions at the secondary
level), f^{-1} exists if and only if f is strictly monotonic on I. This is another way
to characterize functions that admit an inverse function. We look at its proof in the
following section.

8.3   Connecting to Real Analysis

Because most functions in secondary mathematics are continuous and defined on
intervals of real numbers, strict monotonicity is an effective tool in this setting for

determining whether a function has an inverse function and, if not, how to restrict the domain so that it does have one.

8.3.1   Inverse Functions and Strict Monotonicity

Consider the following theorem, mentioned in the previous section, and which will require justification in two directions:

Theorem Suppose I is an interval of $\mathbb{R}$, and $f : I \to \mathbb{R}$ is continuous on I. Then f has an inverse function if and only if f is strictly monotonic.

Proof Let f be a continuous function defined on an interval $I \subseteq \mathbb{R}$. As we work our way through the argument in each direction, keep an eye out for where the assumptions of continuity and a connected domain enter the argument.

In the forward direction, the statement to prove is: "If f has an inverse function then f is strictly monotonic." Here, we consider an indirect proof; we assume f is continuous and possesses an inverse, but that it is not strictly monotonic.

Articulating what it means to *not* be strictly monotonic involves considering several cases. Being strictly increasing or decreasing implies that for all $a < b < c$ in I, it follows that either $f(a) < f(b) < f(c)$ or $f(a) > f(b) > f(c)$. Since f is not strictly monotonic, then there exist $a < b < c$ in I where one of the following two situations must be true:

(i) $f(a) \le f(b)$ and $f(b) \ge f(c)$, or

(ii) $f(a) \ge f(b)$ and $f(b) \le f(c)$.

We prove the first case (i) here—examples of which are depicted in Fig. 8.4. (The proof for the other case follows a similar argument and is requested in Problem 8.9.) Because f has an inverse it must be one-to-one. This means $f(a) < f(b)$ and $f(b) > f(c)$. So our proof here will have to consider two situations. As one situation, consider when $f(c) < f(a) < f(b)$, as depicted in Fig. 8.4a. In this case, we know f is defined and continuous on the interval $[b, c]$. Thus by the Intermediate Value Theorem (IVT), there is an element d in the interval (b, c) such that $f(d) = f(a)$. But since d is in I and $d \ne a$, then f is not one-to-one, which is a contradiction. In the other possible situation, $f(a) < f(c) < f(b)$ depicted in Fig. 8.4b, we can apply the IVT on the interval $[a, b]$ to produce a $d \in (a, b)$ such that $f(d) = f(c)$. Here, again, we reach a contradiction with the fact that f is assumed to have an inverse and thus be one-to-one. We conclude that f must be strictly monotonic on I.

In the reverse direction we need to prove the statement: "If f is strictly monotonic, then f has an inverse function." Choose two arbitrary, but distinct, points, $a \ne b$ in I. Because f is strictly monotonic we know either $f(a) < f(b)$ or $f(a) > f(b)$. In either case this implies $f(a) \ne f(b)$, which means f is one-to-one. Because being one-to-one is sufficient to guarantee an inverse, we are done. □

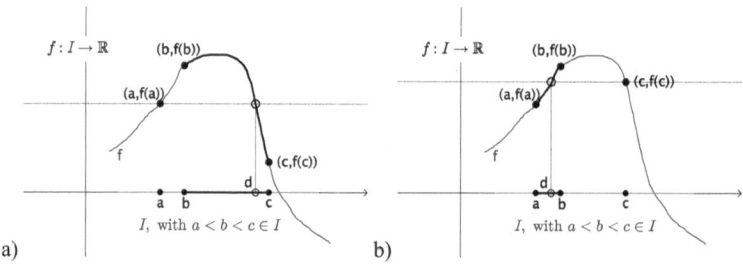

Fig. 8.4 A continuous function f, not strictly monotonic on I, with (**a**) $f(c) < f(a) < f(b)$, and (**b**) $f(a) < f(c) < f(b)$

Note that continuity was not required for our proof in the reverse direction. Can you identify where it was used in the forward direction?

8.3.2 Solving Equations

It's time to take a closer look at what happens when we solve an equation by applying an inverse function to each side, particularly in the case when there are domain restrictions involved in defining the inverse.

Assume we want to solve the equation

$$f(x) = b,$$

where $f : A \to \mathbb{R}$ is a function that is *not* one-to-one, and b is in the range of f. To define an inverse, let A_1 be a subset of the domain A on which f is one-to-one so that f^{-1} exists on A_1. Because b is assumed to be in the range of f, we know there is at least one $x \in A$ satisfying $f(x) = b$. Is $x = f^{-1}(b)$? That depends on whether x is in the restricted domain A_1.

Theorem Suppose $f : A_1 \to \mathbb{R}$ is a function that is one-to-one on A_1, and suppose $b \in f(A_1)$. Then $x = f^{-1}(b)$ is the unique solution to the equation $f(x) = b$ on A_1.

Proof First, we confirm that $x = f^{-1}(b)$ is a solution to the equation $f(x) = b$. Applying f to this choice of x gives $f(x) = f(f^{-1}(b))$. The definition of f^{-1} implies $f(f^{-1}(y)) = y$ for all $y \in f(A_1)$, and so $f(x) = f(f^{-1}(b)) = b$. Hence, $x = f^{-1}(b)$ *is* a solution.

To see that $x = f^{-1}(b)$ is the *unique* solution to the equation in A_1, let x' be an arbitrary solution to the equation $f(x) = b$, meaning we assume $f(x') = b$. Since $b \in f(A_1)$, we can apply f^{-1} to both sides to assert $f^{-1}(f(x')) = f^{-1}(b)$ from which it follows that $x' = f^{-1}(b)$. This proves $x' = x$, confirming that $x = f^{-1}(b)$ is the *only* solution in A_1.   □

This theorem makes explicit what applying the inverse function to both sides of an equation accomplishes. If the function has an inverse on $\mathbb{R}$, applying the inverse function obtains the *unique* solution on $\mathbb{R}$. If the function is not one-to-one on $\mathbb{R}$, this theorem shows that by restricting the function to a smaller domain A_1, applying the inverse function gives the unique solution in A_1 so long as $b \in f(A_1)$. Which means applying the inverse function to both sides of the equation will yield one solution to the original equation, but not necessarily all solutions. Other possible solutions outside A_1 get missed in this process. And how do we recoup these missing solutions? As we saw with the sine function example, properties such as symmetry and periodicity can be leveraged to identify the other solutions to the equation.

8.4   Connecting to Secondary Teaching

The kinds of feedback a teacher might give in the original teaching situation are informed by the mathematical ideas in this chapter. The teacher needs to identify the student's solution as incomplete but, more importantly, the response should draw out key ideas captured by the theorems under discussion. Notably, the student appears to be applying a general "rule" with trigonometric functions: add '$+2\pi k$' at the end of the process. Given the apparent lack of conceptual understanding, the teacher's role is to ensure the student has an accompanying mathematical explanation for that rule (TP.5). Indeed, part of that process involves making explicit some of the assumptions that are left implicit in the solving process.

8.4.1   Strict Monotonicity as a Visual for Domain Restrictions

We've seen how the existence of inverse functions is tied to strict monotonicity and frequently requires restricting a function's domain. This can be challenging for students who might approach mathematics passively. The idea that one can actively change a function by restricting its domain is not so obvious. Students need help overcoming this hurdle. The theorems discussed shift attention from the concept of being one-to-one to strict monotonicity as a means for identifying domain restrictions. Properties such as being one-to-one and passing the "horizontal line test" are typically asked in relation to the function as a *whole*, whereas properties such as strictly increasing or decreasing are often discussed in regard to *parts* of the function (e.g., *where* is the function increasing?). This is potentially useful for helping students recognize their ability to restrict domains since strict monotonicity is inherently "visual." It is possible, based solely on visual inspection, to identify intervals of the domain on which a function that is not strictly monotonic on $\mathbb{R}$ would have an inverse function (see Fig. 8.5a).

In theory, there are many ways to restrict a domain in order to produce an array of partial inverses for a given continuous function. But there is also a convention for defining "the" inverse function. Typically, we look for the largest interval I on which f is strictly monotonic that contains 0 and at least one positive number. Hopefully, $f(I)$ will be the entire range of f as well. Applying this convention to $f(x) = x^2$

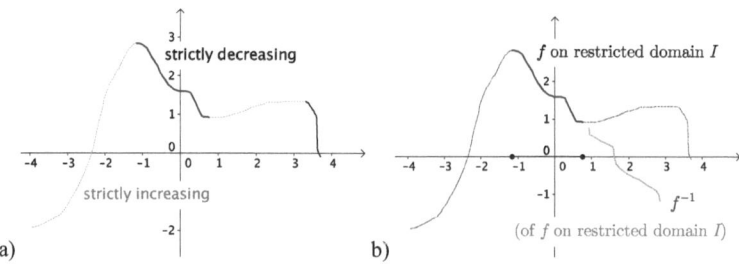

Fig. 8.5 Using strict monotonicity to identify (**a**) domains on which f^{-1} exists, and (**b**) the "conventional" f^{-1}

yields the square root function based on the restricted domain $I = [0, \infty)$. For $f(x) = \sin(x)$, it defines arcsin (y) based on $I = [-\pi/2, \pi/2]$; for $f(x) = \cos(x)$, we get arccos (y) based on $I = [0, \pi]$. In each case, the strict monotonicity over the restricted domain is straightforward to visualize. Finding the conventional inverse for a more unruly function is illustrated in Fig. 8.5b.

8.4.2 Inverses and Missing Solutions

Most steps in the process of solving an equation can be understood as applying an inverse function to both sides. For example, something as simple as subtracting 5 from each side of an equation can be achieved by applying $f^{-1}(x) = x - 5$, which is the inverse of $f(x) = x + 5$. The result, discussed in this chapter, is that the process can end with "missing" solutions when we apply an inverse function whose domain is not $\mathbb{R}$.

This was the case in the original teaching situation, where the student's decision to add '$+2\pi k$' at the end was an attempt to recoup these missing solutions. To make sense of the use of inverse functions in equation solving, and to respond in a way that provides the student with an explanation for their procedures, we need to think carefully about when and why solutions might get "lost" in the process. As a way to be concrete, let's consider the following continuation:

Ms. Gonzalez responds to Meng with the following comments and visuals.

Meng, you did a nice job applying the arcsine to both sides to solve for x, and in adding solutions, '$+2\pi k$', knowing that sine is a periodic function. But let's take a closer look at two of your steps.

In the step

$$\text{(i) } 2x = \arcsin(0.7),$$

(continued)

Ms. Gonzalez uses the illustration below to point out that Meng is finding a single solution, but on a limited part of the graph.

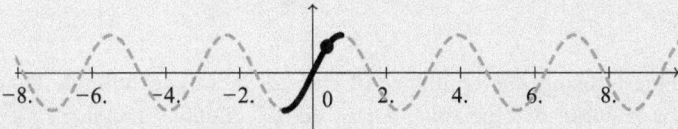

In the step

$$\text{(ii) } x = 0.5 \arcsin (0.7) + 2\pi k,$$

she observes that Meng knows to account for the periodicity of sine curves, and uses the graph below to highlight what Meng is *trying* to accomplish.

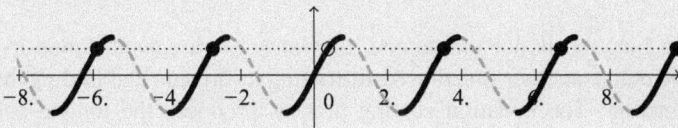

She then highlights the fundamental challenge: the period for repeating solutions in this example is not 2π, but π. She suggests both of the following as possible ways to repair this:

1. When taking the arcsine of both sides you are referring to $\sin (x)$, so the 2π period for repeating angles could be included in *that* step (i):

$$2x = \arcsin (0.7) + 2\pi k$$

   Then, dividing by 2 would result in the correct $+\pi k$.
2. Or, in the *last* step (ii), you need to determine and use the period of $\sin (2x)$ in the original equation, which is π (not 2π) to know how frequently solutions repeat:

$$x = 0.5 \arcsin (0.7) + \pi k$$

From there, Ms. Gonzalez continues, stating that even in this case Meng would *only* have found the solutions identified on the graph above. Solving an equation means finding *all* solutions. Ms. Gonzalez then points to the graph below, indicating students would still need to find "that" solution.

(continued)

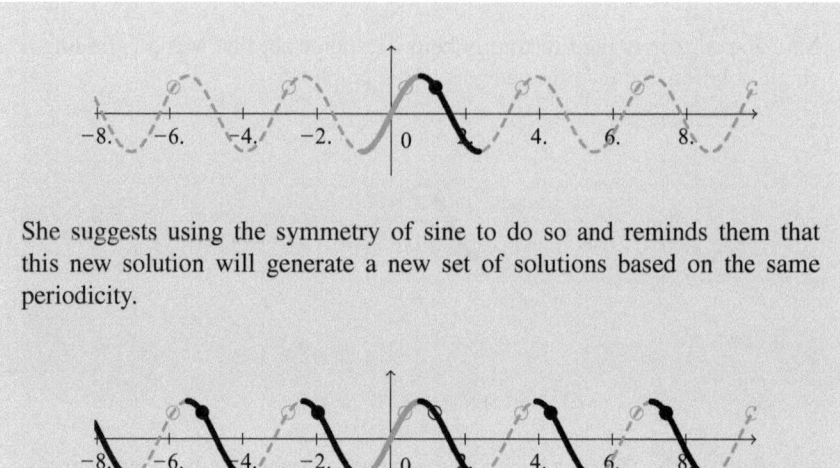

She suggests using the symmetry of sine to do so and reminds them that this new solution will generate a new set of solutions based on the same periodicity.

When a function is not one-to-one, like $\sin(x)$, we cannot define its inverse. Instead, we define another function by restricting its domain and finding the inverse of that function. The equation solving process produces the unique solution in that restricted domain, which means all other solutions must be found by other means. The teacher's response to the student makes these ideas explicit, providing a rationale for why steps like '$+2\pi k$' are included in solving trigonometric equations. These mathematical explanations are important for sense-making, especially when solving is more complex than just "do the same thing to both sides." In the initial step of applying the arcsine function to both sides, the result is valid only on a limited domain; specifically, $2x = \arcsin(0.7)$, when $2x \in [-\pi/2, \pi/2]$. The symmetry of the sine curve means that $2x = \pi - \arcsin(0.7)$ is also a distinct and valid solution. Finally, accounting for the periodicity and solving for x yields the complete list of solutions: $x = 0.5\arcsin(0.7) + \pi k$ and $x = 0.5\pi - 0.5\arcsin(0.7) + \pi k$.

Although in this chapter we have considered a particular equation, similar thinking is needed whenever a function in the initial equation requires a domain restriction to have an inverse function. In solving equations like $(x - 2)^2 = 9$, or $|x - 2| = 9$, the same kinds of methods (e.g., symmetry) can be applied to identify missing solutions. You are asked to do so in several problems at the end of the chapter.

8.4.3 Inverses and Extraneous Solutions

Just as the use of an inverse function can result in missing solutions, it can also lead to extraneous ones. This happens in the reverse context: when we have a function who's inverse function has a "typical" domain that is larger than what

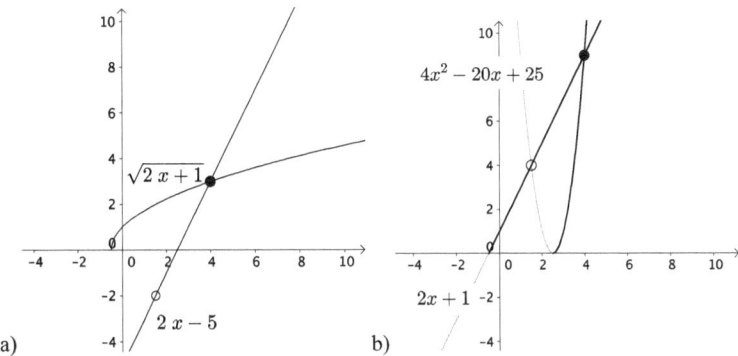

Fig. 8.6 Solving equations and identifying extraneous solutions

the inversion process yields. If we start with $f(x) = \sqrt{x}$, for example, then the inverse is $f^{-1}(y) = y^2$. Because the range of f is $[0, \infty)$, the domain of f^{-1} is considered to be $[0, \infty)$. But the typical domain of the squaring function is all of $\mathbb{R}$. This ambiguity can lead to extraneous solutions.

Let's look at how this issue arises when solving the equation

$$\sqrt{2x + 1} = 2x - 5.$$

Applying the inverse function (squaring) to both sides, we get

$$2x + 1 = 4x^2 - 20x + 25.$$

The graphs of both sets of equations in Figs. 8.6a and b make the extraneous solution evident. This extra solution can be understood based on some inferences from the initial equation: both $2x + 1 \geq 0$ (since we cannot have negative numbers inside the radical) and $2x - 5 \geq 0$ (since the output of the square root function is positive). For the left-hand side of the equation this means $x \geq -0.5$, and for the right-hand side of the equation this means $x \geq 2.5$. We would only consider solutions on which the domains intersect. Another way of stating this would be in relation to the inverse function of $f(x) = \sqrt{x}$, which is $f^{-1}(y) = y^2$ with the restricted domain $[0, \infty)$. Composing on the left-hand side results in $2x + 1$, but only when x is in the domain of $\sqrt{2x + 1}$; i.e., $x \geq -0.5$ (depicted in Fig. 8.6b). On the right-hand side, composing y^2 on $[0, \infty)$ with $2x - 5$, to get $(2x - 5)^2$, will only be valid when the 'y'-output of $2x - 5$ is in the domain of the composing function y^2, which is to say when $2x - 5 \in [0, \infty)$, or $2x - 5 \geq 0$. (Only those produce valid inputs for y^2 when we must have $y \geq 0$). Thus, the composition $4x^2 - 20x + 25$ is valid only on $[2.5, \infty)$. Graphically, this corresponds to the right-half of the parabola shown in Fig. 8.6b.

The key point is that applying the inverse function results in an equation valid only for *certain parts of the line* $2x + 1$ and *certain parts of the parabola* $4x^2 -$

$20x + 25$. Solving $2x + 1 = 4x^2 - 20x + 25$ gives the two solutions $x = 1.5$ and $x = 4$. But we are only interested in solutions where $x \geq 2.5$, meaning $x = 1.5$ is extraneous. Notably, this extra solution is where the line $y = 2x - 5$ in Fig. 8.6a would intersect the reflection of $\sqrt{2x + 1}$ over the x-axis.

Problems

8.1 Consider the graph of $h(x)$ below. (i) Draw the interval I that represents the *conventional domain* on which $h(x)$ has an inverse function. Justify your choice by referencing one of the theorems in the real analysis section. (ii) Explain whether $h(x)$ would have an inverse function on either of the following domains: $[0, \pi/4]$, or $[\pi/2, \pi]$.

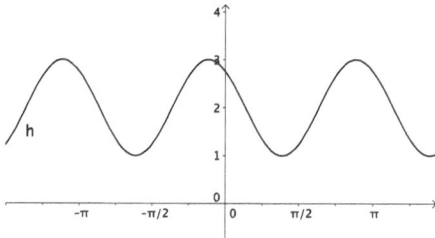

8.2 Solve the following trigonometric function algebraically:

$$2 \cos (x) = \sqrt{2}$$

Make sure to find all solutions. At each step, identify which solutions in the graph (below) you have found.

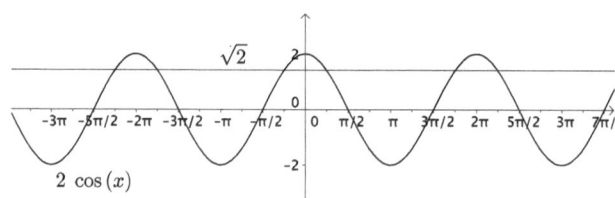

8.3 A student carries out the following solution process:

$$\sin (4x) = -1$$

$$\arcsin (\sin (4x)) = \arcsin (-1)$$

$$4x = \tfrac{3\pi}{2} + 2\pi k$$

$$x = \tfrac{3\pi}{8} + \tfrac{\pi}{2}k$$

As it turns out, the student has found all of the solutions even though they did not carry out all the steps we discussed in this chapter. Describe which step was not done, and explain why the student's approach still found all the solutions, referencing symmetry and periodicity.

8.4 Consider a simple quadratic equation, $x^2 = 25$, and the solution and explanation algebra teachers often provide.

> "We solve this equation by taking the square root of both sides. The square root of 25 is $+5$ and -5. So $x = \pm 5$."

This explanation is mathematically incorrect. By definition, the square root function, $\sqrt{x}$, is the *unique, non-negative number* whose square is x. Thus, $\sqrt{25} = 5$, and $\sqrt{25} \neq -5$. Justify the steps in the equation using ideas from this chapter, in particular, viewing $g(y) = \sqrt{y}$ as the inverse of $f(x) = x^2$, but only on a *restricted domain* of $f(x)$, and using the symmetry of $f(x)$ to find the missing solution.

8.5 Solve the equation $2|x - 5| + 1 = 9$. Explain at what step(s) the symmetrical aspects of the absolute value function are introduced in order to find all solutions to the equation?

8.6 Consider the equation $(x + 1)^2 = 9$. Two possible explanations for solving this equation would be:

1. Think of all real numbers that, when squared, yield 9. These are 3 and -3. So, the next step is to write two equations: $x + 1 = 3$, and $x + 1 = -3$.
2. Whenever you introduce a square root on one side, you need to put a "$\pm$" on the other side. So, the next step is: $\sqrt{(x + 1)^2} = \pm\sqrt{9}$, and so $x + 1 = \pm 3$.

However, many students prefer "doing the *same* thing, to *both* sides" in each step. In this light, option 1 feels less systematic since the equation is split into two equations, and option 2 feels odd since the $\pm$ is added only to one side. A third approach is to consider applying the inverse function of x^2, which can be considered on the restricted domain when $x \in [0, \infty)$, and which is $\sqrt{y}$. When applying the inverse function to both sides, we get $\sqrt{(x+1)^2} = \sqrt{9}$, and so $x + 1 = 3$. Suppose you used this third approach to explain solving the equation. Write out a description of how would you help a student interpret the one solution found, $x = 2$, and how would you help them find the other remaining solution.

8.7 Consider the following algebraic solution that a student submitted:

$$-\sqrt{x-3}+2=3$$

$$-\sqrt{x-3}=1$$

$$\left(-\sqrt{x-3}\right)^2=(1)^2$$

$$x-3=1$$

$$x=4$$

Use the graph and the inverse function of $-\sqrt{x}$, which is y^2 on $(-\infty, 0]$, to explain why the solution $x = 4$ is an extraneous solution. This would mean there are no solutions to the original equation.

8.8 Consider the function in the previous Problem (8.7), $h(x) = -\sqrt{x-3}+2$. The problem amounts to solving $h(x) = 3$. At some point in the solving process from the previous problem, the student squared both sides. More specifically, they applied y^2 to both sides, where y^2 on $(-\infty, 0]$ is the inverse function of a *different* function, $-\sqrt{x}$. i) Find the inverse function of $h(x)$. (Make sure to think about any domain restrictions.) ii) Apply h^{-1} to both sides of the equation $h(x) = 3$ to solve the equation.

8.9 The proof of the theorem in Sect. 8.3.1 in this text considered case (i): $f(a) < f(b)$ and $f(b) > f(c)$. The proof for case (ii), $f(a) > f(b)$ and $f(b) < f(c)$, was not given. It amounts to situations in which $f(b) < f(a) < f(c)$, or $f(b) < f(c) < f(a)$. Sketch a function f that would fit each of these second two cases—as per the conditions, f should be continuous on an interval I, and with $a < b < c$ in I. Then, provide a proof for case (ii), which should cover both of these second two cases. It will be similar to the proof for case (i) in the text.

References

1. Abbott, S. (2015). *Understanding analysis* (2nd ed.). New York, NY: Springer.

Differentiability and the Secant Slope Function 9

9.1   Statement of the Teaching Problem

The familiar concept of the *slope* of a line can be understood from multiple points of view. Computationally it is given by the formula $\frac{y_2-y_1}{x_2-x_1}$, where (x_1, y_1) and (x_2, y_2) are two points on the line. Geometrically it represents how much the y-value increases or decreases for each 1-unit increase in the x-value. Physically the slope represents the rate of change of y with respect to x. For instance, if y is the height of an object (in meters) at time x (measured in seconds), then the slope is the velocity of the object measured in meters per second.

The core idea of differential calculus is to generalize the fundamental notion of slope to functions that are not straight lines. With linear functions, the slope is a *global* feature—it is constant and can be computed using any two points on the line. For non-linear functions we have to shift our understanding of slope from a global feature to a *local* one. We ask, "What is the slope, or rate of change, at a particular point on the function?" This is the impetus behind the concept of the derivative. Like the notion of slope, the derivative can be viewed from multiple points of view. It has a formal definition that incorporates the computational formula of slope, which we review momentarily. Geometrically, the derivative can be understood as the "slope" as a point on a curve. Likewise, the physical interpretation of the derivative is in regard to a rate of change, but there is the significant complication in that this rate of change can be different at different points on the function. This is what it means to say the slope is a local feature. First we must specify a point on the function, and then we set ourselves the task of finding the "slope" at this particular point.

This brings us face to face with the central dilemma of defining the derivative. Intuitively, the derivative is the "slope" at a single point on the graph, but the familiar algebraic formula for computing slope requires two points.

Consider the following pedagogical situation:

A calculus teacher, Mr. Petrov, is having students practice calculating the derivative and uses a problem from the textbook:

$$\text{Calculate } f'(1) \text{ for } f(x) = \begin{cases} 2x + 1, \ x \geq 1 \\ 2x - 1, \ x < 1 \end{cases}$$

After some discussion, the class ends up split into two groups claiming two different answers. The first side argues $f'(1) = 2$ because derivatives are about slope, and the slope is 2 for both parts of this piecewise definition. The second side argues $f'(1)$ is not defined, because they remember something about differentiable functions needing to be continuous, and the function is discontinuous at $x = 1$.

Although both groups give justifications for their conclusions, both cannot be correct. These kinds of disagreements are common in teaching. Identifying which group has the right answer is a necessary part of the teacher's responsibility; but the more important challenge is finding a justification that explains why the conclusion is valid (TP.5). In situations where students cannot agree, both sides likely have something meaningful—yet incomplete—to contribute to the conversation. In this scenario, slope and continuity are both connected to the derivative, and each concept turns out to be essential for finding a satisfying resolution to the debate.

Before reading on, which group do you think has come to the right conclusion? Do you feel their reasoning is sufficient? If you were the teacher, how would you respond to the students in the class?

9.2   Connecting to Secondary Mathematics

9.2.1   Problematizing Teaching and the Pedagogical Situation

Let's start with a closer look at how each side arrived at its conclusion.

The group with the answer $f'(1) = 2$ is evidently thinking about the derivative as slope. Vertically-translated lines like $y = 2x$, $y = 2x + 1$, and $y = 2x + 2$ all have the same slope. Interpreting the derivative as the slope of the curve, this group reasoned that even though the function has a discontinuity—it "jumps" up— the slope does not change with this vertical shift and hence the derivative should not change. This conclusion emerges naturally from looking at each piece of the function individually. If we define the first part of f to be $f_1(x) = 2x + 1$ for $x \geq 1$, then it would appear that $f_1'(x) = 2$ for $x \geq 1$ (and at least certainly for $x > 1$). Likewise, if we define the second part to be $f_2(x) = 2x - 1$ for $x < 1$, then its derivative is clearly $f_2'(x) = 2$ for $x < 1$. Since approaching $x = 1$ from the right side and the left side both point to the conclusion $f'(1) = 2$, there is a compelling case to be made that $f'(1)$ should indeed equal 2.

The second group concludes the derivative is not defined and appears to be reasoning based on a theorem about derivatives. Their argument relies on the recollection that differentiability requires continuity (Diff. $\implies$ Cont.) or, in logical terms, "If a function is differentiable at a point then it is continuous at that point." The group is actually invoking the contrapositive of this result: "If a function is not continuous at a point, it is not differentiable either." When working with directional statements like these, it can be easy to confuse the contrapositive with its converse, and there is also the possibility that the group has misremembered the theorem altogether. Perhaps it is that continuous functions are differentiable (Cont. $\implies$ Diff.)? When relationships are memorized, the chain of reasoning can only go so far. Regardless of whether they have the right answer, the students have not provided any further reasoning to support their conclusion, which suggests their thinking is based solely on a recollection of some rule and not a robust understanding of how continuity impacts the existence of the derivative.

Pause to consider whether either of these two answers seem more correct than the other now. Was there anything that changed your mind?

9.2.2   Recognizing Computations as Singular Objects

Many mathematical ideas start out as processes, and then those processes turn into objects of study themselves. Some suggest learning mathematics is akin to development through such conceptual transitions.[1] A computation is initially understood as a procedure on constituent *parts*, but when we start reasoning about the resultant computation as a *whole* without actually executing the procedure, we are recognizing the computation as a singular object.[2]

Consider a basic computation like $2 + 5$. Children first learn about addition in terms of physical actions. They might put a group of 2 marbles together with a group of 5 marbles and then count all the marbles—so the answer is 7. The symbol '+' signifies that two groups of objects should be joined together, defining addition as the process of joining *two individual numbers*, $(2) + (5)$. But as our understanding of addition progresses, there comes a moment when the computation becomes a *singular object* itself—the "sum." We do not need to go through the process to arrive at 7; instead we regard the entire calculation as a singular expression, $(2+5)$. The calculation is understood as an object, interchangeable with 7 because they both represent the sum.

Another example of this phenomenon, especially pertinent to discussing the derivative, is the computation of slope. Initially, the formula $\frac{y_2-y_1}{x_2-x_1}$ is interpreted

---

[1] APOS theory [2] posits that mathematical learning transitions through phases; understanding mathematical concepts as actions (A), then processes (P), then objects (O), then schemas (S).

[2] Sfard [3] describes this process as "reification."

by students as a series of three computations—calculate $\Delta y = y_2 - y_1$ and $\Delta x = x_2 - x_1$, and then divide. But eventually it is important to see the entire expression $\left(\frac{y_2 - y_1}{x_2 - x_1}\right)$ holistically as representing the slope of the line through two given points.

9.3   Connecting to Real Analysis

To connect this discussion to the content of real analysis content, let's look carefully at the formal definition of the derivative and appreciate how it incorporates the notion of slope. For a given function $g(x)$ and a given value c in the domain of g, the derivative is formally defined as

$$g'(c) = \lim_{x \to c} \frac{g(x) - g(c)}{x - c}.$$

The first critical step in understanding this definition is recognizing the quotient $\frac{g(x) - g(c)}{x - c}$ as the slope calculation between two points—a given point $(c, g(c))$ and another point $(x, g(x))$ somewhere else on the function. The second step is appreciating the role of the limiting operation $\lim_{x \to c}$. This limit really asks, "What happens to the slope value as the point $(x, g(x))$ becomes 'very close' to the fixed point $(c, g(c))$?" This is how calculus solves the riddle of computing the slope at a single point on a curve. The idea is to compute the old-fashioned slope through two points on the curve and then think about what happens as those two points get closer together. This is where the tools of real analysis such as limits, and whether they exist, can be brought to bear.

9.3.1   The Secant Slope Function

Throughout the subsequent sections, we will use the example

$$g(x) = \tfrac{1}{10}x^3 - \tfrac{3}{10}x^2 - \tfrac{9}{5}x + 5.$$

Like the elementary school student learning about addition for the first time, it is natural at first to view the definition of $g'(c)$ as a computational process involving constituent parts. First, we must pick a fixed value, c. Let's start with $c = 2$ as an example, which means we can compute $g(c) = g(2) = 1$. So both c and $g(c)$ are real numbers. To compute $\frac{g(x) - g(c)}{x - c}$, we might give ourselves a selection of x values. Taking $x = 1$, for instance, yields $g(1) = 3$. This gives us two points on the graph of g—the fixed point $(2, 1)$, at which we are trying to find the derivative, and a second point $(1, 3)$. The slope through these two points is $\frac{3-1}{1-2} = \frac{2}{-1} = -2$. Table 9.1 displays the results of repeating this calculation for a selection of other x-

Table 9.1 Computing the slopes of individual secant lines with $c = 2$

x	$(x, g(x))$	$(2, g(2))$	$g(x) - g(2)$	$x - 2$	$\frac{g(x)-g(2)}{x-2}$
1	(1,3)	(2,1)	2	−1	−2
1.5	(1.5, 1.96...)	(2,1)	0.96...	−0.5	−1.92...
2	(2,1)	(2,1)	0	0	und.
2.5	(2.5, 0.18...)	(2,1)	−0.81...	0.5	−1.62...
3	(3, −0.4)	(2,1)	−1.4	1	−1.4
4					
5					

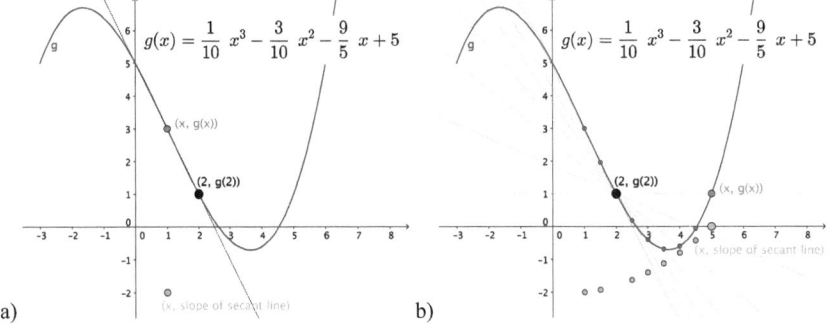

Fig. 9.1 Secant line(s) and slope(s), with $c = 2$, for a) $x = 1$, and b) $x = 1, 1.5, 2, 2.5, 3, 3.5, 4, 4.5,$ and 5

values. Note that for $x = 2$ the calculation breaks down. (Do you see why?) Before moving on, complete the table for $x = 4$ and $x = 5$ (while keeping $c = 2$).

A line through two points on a curve is called a *secant line*. The secant line through $(1, 3)$ and $(2, 1)$ on g is depicted in Fig. 9.1a. This corresponds to our using $x = 1$ and results in a secant line with slope -2. To capture this result, Fig. 9.1a also contains the point $(1, -2)$. Figure 9.1b plots all of the secant lines associated with the x-values in the table, as well as the points corresponding to the secant line slopes for each value of x. Turning our attention to the definition of $g'(2)$, we see that we are most interested in computations when the x-values are close to $c = 2$. At $x = 1.5$ we found a slope of approximately -1.92; at $x = 2.5$ the slope is roughly -1.62. These values give us a sense of the slope as we approach $c = 2$.

This process of computing individual slopes is a useful way to begin to understand the definition of the derivative. However, focusing too intently on individual computations does not capture the big picture. A proper understanding requires looking holistically at this process across all x-values, secant lines, and slopes. Instead of interpreting $\frac{g(x)-g(c)}{x-c}$ as a series of arithmetic computations, it is more helpful to view the expression as a singular object. Notably, since $g(x)$ and $x - c$ are functions, the expression is a quotient of two functions, and so the "object" it represents is a (new) *function*. This means thinking about $\left(\frac{g(x)-g(c)}{x-c} \right)$ as a function

Fig. 9.2 The secant slope
function of g with $c = 2$

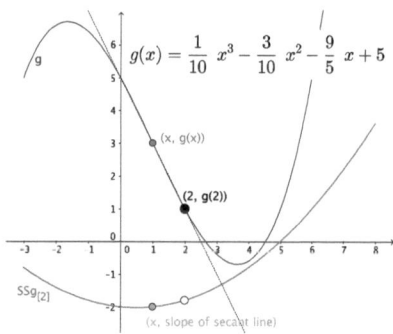

of x.[3] The case where $c = 2$ which we are currently considering yields the function $\left(\frac{g(x)-1}{x-2}\right)$. What is this new function? The x inputs are the locations of the second point $(x, g(x))$ (the first point for us is always $(2, 1)$), and the outputs are the slope values of the resulting secant lines. We refer to as this new function as the *secant slope function*. Its' construction is specific to a given point c and function $g(x)$; the notation we use to denote it, $SSg_{[c]}(x)$, specify both of these givens.

Definition For a given function $g(x)$, and a given point c (in the domain of g), the **Secant Slope Function** $SSg_{[c]}(x)$ is $\frac{g(x)-g(c)}{x-c}$, which takes each input x (in the domain of g), and outputs the slope of the secant line between $(x, g(x))$ and $(c, g(c))$.

Figure 9.2 depicts the graph of the secant slope function for our example with $c = 2$, which is $SSg_{[2]}(x) = \frac{g(x)-1}{x-2}$. Comparing Figures 9.2 and 9.1b highlights the fact that the secant slope function is the composite result of *all* of the individual calculations from different x-values. Indeed, since x can vary, we might even imagine the second point on the function "moving," and at each new x-value representing the slope of the corresponding secant line.

We are now at a better vantage point from which to think about the 'lim$_{x \to c}$' portion of the definition of $g'(c)$. From the more primitive computational point of view, the best we can do is gather empirical evidence about the limit from a handful of slope computations when x is close to c. Transitioning to the idea that $\frac{g(x)-g(c)}{x-c}$ is not a multi-step computational process but a singular function provides new insight. In this light, the derivative at c is understood as the functional limit as x approaches c of this new secant slope function; that is, $g'(c) = \lim_{x \to c} SSg_{[c]}(x)$. Even without engaging the formal $\varepsilon - \delta$ definition of functional limits, it is more straightforward to make a determination on whether the limit of the secant slope function exists as x approaches c. Every $SSg_{[c]}(x)$ has a hole when $x = c$ (why?), but now we see that

---

[3] It is helpful to remind yourself c is a constant in this expression (even though we will also eventually want to think of many different c-values).

Fig. 9.3 Graph of $SSg_{[-1]}(x)$, where $g'(-1)$ is found by evaluating '$\lim_{x \to -1}$' of this function

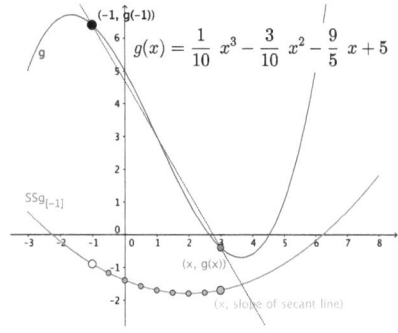

$g'(c)$ is equivalent to whatever value is the functional limit $\lim_{x \to c} SSg_{[c]}(x)$, if it exists, which is clearer to conceptualize. In the case of our particular example, this limit on our secant slope function comes out to be -1.8.

9.3.2 The Derivative as a Function

Having reached a more robust understanding of the definition of the derivative at a given point c in the domain of g, we make one last conceptual leap and consider c to be a variable so that $g'(c)$ becomes a function in its own right. For $c = 2$ we found $g'(2) = -1.8$. To find $g'(c)$ for some other value of c we return to the secant slope function $SSg_{[c]}(x)$ and note that this function will be different for different values of c. Figure 9.3 demonstrates this difference; it shows the secant slope function for the same function g but with a new value $c = -1$. To compute $g'(-1)$ we take the limit of $SSg_{[-1]}(x)$ as x approaches -1 to find $g'(-1) = -0.9$.

In general, each value of c determines a specific secant slope function via the formula

$$SSg_{[c]}(x) = \frac{g(x) - g(c)}{x - c},$$

from which we compute $g'(c)$ by taking the limit as x approaches c. The derivative function g' is then a record or compilation of *all* these functional limits obtained from *all* of these different secant slope functions. Table 9.2 shows the computations for generating the derivative function at a few selected values of c, and Fig. 9.4 depicts g'. The figure also displays a particular point $(c, g'(c))$ on the derivative function, which is the result of evaluating $\lim_{x \to c} SSg_{[c]}(x)$. To illustrate this relationship, the secant slope function $SSg_{[c]}$ for this singular value of $c = 2$ is also included.

Before moving on, try to replicate the processes in this section until you feel comfortable with the idea of a secant slope function. Using $g(x) = x^2$ as an example, set $c = 1$ and sketch the secant lines you get using $x = 3, 2, 2.5$ and 2.1. Find an algebraic expression for $SSg_{[1]}(x)$ and simplify as much as possible.

Table 9.2 Computations for $g'(x)$ from secant slope functions, for $x = 2, 1, 0, -1$

c	$SSg_{[c]}(x)$	$\lim_{x \to c}$	$g'(c)$	$(x, g'(x))$
2	$SSg_{[2]}(x) = \frac{g(x)-1}{x-2}$	$\lim_{x \to 2} SSg_{[2]}(x) = -1.8$	-1.8	$(2, -1.8)$
1	$SSg_{[1]}(x) = \frac{g(x)-3}{x-1}$	$\lim_{x \to 1} SSg_{[1]}(x) = -2.1$	-2.1	$(1, -2.1)$
0	$SSg_{[0]}(x) = \frac{g(x)-5}{x-0}$	$\lim_{x \to 0} SSg_{[0]}(x) = -1.8$	-1.8	$(0, -1.8)$
-1	$SSg_{[-1]}(x) = \frac{g(x)-6.4}{x+1}$	$\lim_{x \to -1} SSg_{[-1]}(x) = -0.9$	-0.9	$(-1, -0.9)$

Fig. 9.4 Graph of $g'(x)$; the point labeled $(c, g'(c))$ is the $\lim_{x \to c} SSg_{[c]}(x)$ (for $c = 2$)

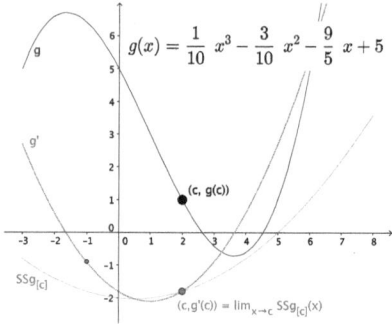

$$g(x) = \frac{1}{10}x^3 - \frac{3}{10}x^2 - \frac{9}{5}x + 5$$

$(c, g'(c)) = \lim_{x \to c} SSg_{[c]}(x)$

What is $\lim_{x \to 1} SSg_{[1]}(x)$? Try this again for $c = 0$. Find $SSg_{[c]}(x)$ for an arbitrary c and use it to find $g'(c)$. How about $SSg_{[c]}(x)$ for $g(x) = x^3$? You should be very comfortable with secant slope functions before reading on!

9.3.3 Derivatives and Continuity

After defining the derivative at a point, it is common in a real analysis course to state and prove the following result (Theorem 5.2.3 in Abbott's [1] text).

Theorem If g is differentiable at a point c, then g is continuous at c as well.

Proof By assumption, $g'(c)$ exists, meaning the $\lim_{x \to c} \frac{g(x)-g(c)}{x-c}$ exists. Based on the Algebraic Limit Theorem for functional limits, we have:

$$\lim_{x \to c} (g(x) - g(c)) = \lim_{x \to c} \left(\frac{g(x) - g(c)}{x - c} \right) (x - c)$$

$$= \lim_{x \to c} \left(\frac{g(x) - g(c)}{x - c} \right) \cdot \lim_{x \to c} (x - c)$$

$$= g'(c) \cdot 0$$

$$= 0.$$

It follows that $\lim_{x \to c}(g(x) - g(c)) = 0$, and thus $\lim_{x \to c} g(x) = g(c)$. This shows g is continuous at c.   □

The proof just given is a standard one for this theorem (it's the one in Abbott, for instance), but it engages in a bit of algebraic trickery. Beginning with the expression $g(x) - g(c)$, we first multiply by 1 written in the form $\frac{x-c}{x-c}$. (We don't have to worry about the case where $x - c = 0$ because the limit as x approaches c is independent of what happens when $x = c$.) The rest of the proof is a straightforward application of the Algebraic Limit Theorem for functional limits (Corollary 4.2.4 in Abbott). Although it is concise and properly substantiates the claim, this proof does not offer much intuition for *why* the theorem is true.

In search of a more enlightening proof, let's consider the logically equivalent contrapositive statement:

Corollary If g is not continuous at a point c, then g is not differentiable at c either.

This statement goes to the heart of the teaching challenge described at the beginning of the chapter. Finding an intuitively appealing argument for this corollary would not only help resolve the debate from the classroom scenario, but it would provide the basis for a new proof of the original theorem. Secant slope functions provide just such a conceptual explanation for why functions are not differentiable at points where they are not continuous.

In his textbook, Abbott [1, p. 142] explains that discontinuities fall into three categories: (i) *removable* discontinuities; (ii) *jump* discontinuities; and (iii) *essential* discontinuities. Let's consider the case of a removable discontinuity. (In Problems 9.6 and 9.7 you will look at the other two categories.) Consider the following function which has a removable discontinuity at $c = 2$:

$$f(x) = \begin{cases} x + 1, & x \neq 2 \\ 1, & x = 2 \end{cases}.$$

In Fig. 9.5, a particular secant line, as well as the secant slope function $SSf_{[2]}(x)$, is depicted. As you look at the figure, make sure you understand the values being plotted by the secant slope function! Remember that the removable discontinuity— the point $(c, f(c)) = (2, 1)$—is the fixed point in constructing secant lines, and $f'(2)$ is defined to be the limit as $x \to 2$ of the constructed secant slope function. In the figure we can see the limit of that secant slope function *does not exist* as we approach 2. Because the point of discontinuity at $(2, 1)$ lies on every every secant line, the lines have slopes that tend toward infinity when the second point x gets closer to 2. From the right the slopes tend toward positive infinity; from the left they tend toward negative infinity. These are reflected in the graph of $SSf_{[2]}(x)$. Either of these conditions is enough to conclude that $\lim_{x \to 2} SSf_{[2]}(x)$ does not exist and so $f'(2)$ does not either.

Fig. 9.5 $SSf_{[2]}(x)$ for a function f with a removable discontinuity at $c = 2$

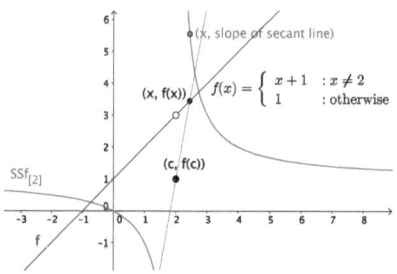

The existence of $f'(c)$ depends on $SSf_{[c]}(x)$ being well-behaved as x approaches c. What this example reveals is that a discontinuity at c creates the possibility for $SSf_{[c]}(x)$—i.e., the slopes $\frac{f(x)-f(c)}{x-c}$—to get unboundedly large as x gets closer to c. With values of $\frac{f(x)-f(c)}{x-c}$ heading off to infinity, there is no way for $\lim_{x\to c} SSf_{[c]}(x)$ to exist, which by definition means $f'(c)$ is not defined at that value.

9.4   Connecting to Secondary Teaching

In the initial teaching situation, students split into two dissenting factions about how to determine the derivative at a point of discontinuity. Results from the previous section show that the correct answer is the derivative does not exist, but we still need to consider how to help the class as a whole come to agreement and understand why this answer is correct. Simply stating that differentiable functions must be continuous and hence the derivative is undefined runs contrary to TP.5—we would have given a rule without giving any kind of mathematical justification. Providing the standard proof for the theorem that differentiability implies continuity might convince a few students that the answer is correct, but still not help them conceptually understand what's really happening.

9.4.1   Navigating Disagreement in the Classroom

Sometimes all it takes to resolve a disagreement is to have students look again at the problem in order to clarify what is being asked or to identify a small error. Revisiting the definition for a central concept is also a reasonable heuristic when there is disagreement; doing so affords an opportunity to resolve potential misconceptions. But definitions can be sufficiently complex that coordinating the parts along with the whole is difficult. This challenge was illustrated in this chapter by having to view the quotient $\frac{g(x)-g(c)}{x-c}$ not only as a calculation for slope, but as a holistic function as well.

In the teaching situation from this chapter, both groups of students drew on important ideas about the derivative and neither could convince the other group of its conclusion. Here is a potential continuation of the scenario that highlights

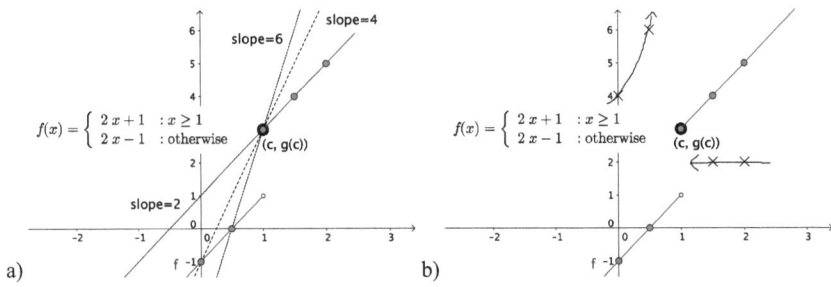

Fig. 9.6 For the function f, a) secant lines and slopes with $c = 1$, and b) a sketch of $SSf_{[1]}$

some useful practices for navigating disagreements and providing meaningful justifications.

Mr. Petrov plots the function $f(x) = \begin{cases} 2x + 1, \ x \geq 1 \\ 2x - 1, \ x < 1 \end{cases}$ on the board, and rewrites the formal definition of $f'(c)$. He circles the point $(1, 3)$ on f, and asks four students to come up to the board:

> For each x-value, one of you please plot the point on the function: at $x = 2$, $x = 1.5$, $x = 0.5$, and $x = 0$. Then, draw the secant line connecting the circled point at $(1, 3)$ to your other point on f.

After the students mark the secant lines (depicted in Fig. 9.6a), Mr. Petrov asks the class to give an approximate value for the slope of each secant line. They give correct slopes for each: 2, 2, 6, and 4 respectively.

He then marks an 'X' at the corresponding points: $(2, 2)$, $(1.5, 2)$, $(0.5, 6)$, and $(0, 4)$ (see Fig. 9.6b). Sketching lines through the Xs, he says:

> Slope is a key idea for determining the derivative. From the right-hand side, the slopes of the secant lines are always 2, but from the left-hand side the secant lines get steeper and steeper because one of the two points is *always* $(1, 3)$. According to the definition, to determine $f'(1)$ we need to look at *this* (sketched) function and evaluate the limit at $x = 1$. Notice the limit does not exist. This is why the derivative is not defined at the point of discontinuity.

With the goal of having both groups of students come to a shared understanding, Mr. Petrov affirms parts of both arguments. In disagreements, each side often has valid but incomplete points. This means student thinking can be leveraged by endorsing some aspects of their arguments while also still pushing for additional refinement. Mr. Petrov confirms the importance of slope for the first camp, asking students to plot lines and identify their slopes. Notably, however, they are slopes of secant lines rather than the sought after slope of the curve at $c = 2$. He also reinforces the observation that the secant line slopes are 2 on the right-hand side of the function before pointing out the problem that arises on the left. With regard to the second camp, the teacher does not simply state the theorem the group recalled. Rather than a rule, Mr. Petrov crafts a hands-on explanation for why $f'(2)$ does not exist at that point of discontinuity that entails a thoughtful engagement with the limit definition of the derivative. Without using the term "secant slope function," Mr. Petrov employs this very concept to push the students to think more deeply about how continuity is related to the existence of the derivative.

Incorporating the notion of a secant slope function with respect to a debate about the derivative does more than just substantiate a particular claim—it affords a conceptual explanation for *why* the claim is true. In Mr. Petrov's response, the sketched secant slope function provides the intuition for why the derivative cannot exist at a point of discontinuity. Anchoring one point of the secant line to a point of discontinuity leads to the recognition that the secant slope function is unbounded in any neighborhood of the discontinuity; concrete "jumps" in y-values, over arbitrarily small x-values, mean the slope values will tend to infinity. More so than with algebraic arguments, the visual impact of steeper and steeper secant lines, together with an unbounded secant slope function, provides a powerful and, often, preferred source of insight for learners, helping meet the expectations of TP.5.

Problems

9.1 Draw the graph of the indicated secant slope function, explain why the secant slope function is not defined at the given c value, and explain how to use the graphs of the secant slopes functions to justify the derivative value at c even though the secant slope function is not defined there.

a) Draw $SSf_{[1]}(x)$ for $f(x) = 3x$ (i.e., with $c = 1$).
b) Draw $SSg_{[1]}(x)$ for $g(x) = x^2$ (i.e., with $c = 1$).
c) Draw $SSh_{[0]}(x)$ for $h(x) = \sin(x)$ (i.e., with $c = 0$).

9.2 In the example function used in this chapter, $g(x) = \frac{1}{10}x^3 - \frac{3}{10}x^2 - \frac{9}{5}x + 5$, we stated that the secant slope function when $c = 2$ was given by $\frac{g(x)-g(2)}{x-2}$. Algebraically, make the substitutions and express $SSg_{[2]}(x)$ as a rational function. Confirm for each input x that this function outputs the slope of the secant line connecting $(x, g(x))$ with $(2, 1)$. You might also try to factor the numerator to confirm there is indeed a removable discontinuity at $x = 2$.

9.3 On a linear function $f(x) = mx + b$, demonstrate that for a given point $(c, f(c))$ the secant slope function is the horizontal line $y = m$ (but with a hole at $x = c$). You might approach this both graphically, sketching several secant lines, and algebraically. Use this fact to justify that the functional limit of each $SSf_{[c]}(x)$ as $x \to c$ is always the constant m.

9.4 A common refrain in calculus is the idea that all "smooth curves" are differentiable. Consider the function $f(x) = x^{1/3}$, which is a "smooth curve" and the point $(0, 0)$ on that curve. First, sketch the secant slope function for $c = 0$, which is $SSf_{[0]}(x)$. Second, explain why the limit of this secant slope function does not exist as x approaches 0, which means $f'(0)$ is undefined. Third, TP.1 suggests that we acknowledge and revisit limitations in such statements. Describe how you would talk about how and when "smooth curves" are differentiable with a class of calculus students. Provide an explanation and any corresponding visuals that you would use to help a group of students understand when and why a point on a "smooth" curve might not be differentiable.

9.5 Consider another function with a *removable discontinuity*. Suppose

$$f(x) = \begin{cases} 3 \text{ if} x \neq 1 \\ 4 \text{ if} x = 1 \end{cases}.$$

Use a secant slope function to present a graphical argument, in connection with the definition of derivative, for why $f(x)$ is not differentiable at $x = 1$—one you might be able to use while teaching a calculus class.

9.6 Consider a function with a *jump discontinuity*. Suppose

$$g(x) = \begin{cases} 1 \text{ if} x \leq 2 \\ 3 \text{ if} x > 2 \end{cases}.$$

Use a secant slope function to present a graphical argument, in connection with the definition of derivative, for why $g(x)$ is not differentiable at $x = 2$—one you might be able to use while teaching a calculus class.

9.7 Consider a function with an *essential discontinuity*. Suppose

$$h(x) = \begin{cases} \frac{1}{(x-2)^2} & \text{if} x \neq 2 \\ 1 & \text{if} x = 2 \end{cases}.$$

Use a secant slope function to present a graphical argument, in connection with the definition of derivative, for why $h(x)$ is not differentiable at $x = 2$—one you might be able to use while teaching a calculus class.

9.8 We often talk about differentiability in terms of a curve being smooth; meaning that functions are not differentiable at "sharp" points. Consider a function f that has a sharp point at $x = c$ (such as $f(x) = |x|$). Sketch out a general argument (i.e., don't limit yourself to the absolute value function!) for why a function with a sharp point at $x = c$ would not have a derivative at that point. Draw on the secant slope function at that point in your explanation. Discuss how you would talk about differentiability at sharp points with a class of calculus students. If you find it helpful, you can consider a sharp point as being where two lines intersect (although there are also other types of "sharp" points).

9.9 Consider the following piece-wise defined function: $f(x) = \begin{cases} x^2, & x \in \mathbb{Q} \\ 0, & x \notin \mathbb{Q} \end{cases}$.
After sketching out this function, which group from the original pedagogical situation—the "slope" group, or the "discontinuity" group—might argue that $f'(0)$ is undefined, and which might argue that $f'(0) = 0$? Explain your reasoning. Now, determine $f'(0)$ from the definition. You might sketch out the associated secant slope function to help determine whether the relevant limit exists and, if so, its value.

9.10 In introducing the derivative in Section 5.1, Abbott states:

A particularly useful class of examples for this discussion are functions of the form:

$$g_n(x) = \begin{cases} x^n \sin(1/x) & \text{if } x \neq 0 \\ 0 & \text{if } x = 0 \end{cases}.$$

What teaching principle do you think this illustrates? To what "discussion" is Abbott referring—meaning, for what mathematical point about derivatives is this class of examples useful?

Turning the Tables

Reflecting on *teaching* from your *learning* in real analysis: TP.1

As another opportunity to reflect on teaching, we explore an aspect of learning real analysis that exemplifies another one of our teaching principles.

TP.1 is about acknowledging and revisiting assumptions and mathematical limitations in mathematics teaching and learning. We use one exercise in Abbott's textbook to ground the discussion. Exercise 5.2.10 (p. 154) states:

> A familiar mantra from calculus is that a differentiable function is increasing if its derivative is positive, but this statement requires some sharpening in order to be completely accurate.

In this exercise, Abbott gives what we will refer to as a "Calculus 1" mantra: *a differentiable function is increasing if its derivative is positive*. He then suggests this statement is not completely accurate; that there are limitations in the statement that should be acknowledged. The exercise asks students to probe the statement by considering the function, $g(x) = \begin{cases} x/2 + x^2 \sin(1/x) & \text{if } x \neq 0 \\ 0 & \text{if } x = 0 \end{cases}$. Essentially, the function $g(x)$ provides a situation for which a point on a differentiable function, $(0, 0)$, has a positive derivative at 0, but where the function is not increasing on any interval that includes 0—meaning, the Calculus 1 mantra needs some work. Abbott then notes in the exercise, if $f'(x) > 0$ for all x in an interval (a, b) (and not just for *one* point), then it is true that the function will be increasing on that interval. That is, as stated, the Calculus 1 mantra is not quite correct (even though it is helpful), but it would be correct with a slight modification (a modification that's probably beyond what most students in Calculus 1 are ready for).

We elaborate a few points related to TP.1. First, special examples, like the function g in Abbott's exercise, can be used to illustrate mathematical constraints or limitations. In Chap. 2 we discussed boundary examples, ones that probe the "edges" of a concept. The current chapter included a number of possible boundary examples: the jump discontinuity in the pedagogical situation; the function $f(x) = x^{\frac{1}{3}}$ at $x = 0$ in one of the homework problems (9.4); and now Abbot's function $g(x)$ in the section above. In each case, the example illustrated the "edges" of behavior, and made a statement's limitations evident. Such examples are one way that TP.1 might be connected to TP.2. Second, limited statements still often contain kernels of truth. Here, the Calculus 1 mantra simply needed to be adjusted to ensure we are talking about the derivative being positive on all points in an interval—not just a single one. The point being that an extra clarification, or additional assumption, can often turn statements with limitations into more accurate ones. Third, enacting TP.1 in the classroom does not mean all statements be perfectly accurate but rather, at some point, that the

(continued)

limitations be discussed. Teachers need to communicate to students in ways they can understand, which often requires simplification—the opposite of precision. To be precise, we include all qualifications and assumptions; to be simple, we give broad descriptions and analogies. TP.1 suggests teaching need not be only one or the other. Derivatives should be talked about as slopes; they should be connected to a function being increasing; but the story cannot end there. Precisely "when" limitations are acknowledged is less of the point than "that" they are acknowledged.

References

1. Abbott, S. (2015). *Understanding analysis* (2nd ed.). New York, NY: Springer.
2. Dubinsky, E. (2014). Actions, Processes, Objects, Schemas (APOS) in mathematics education. In S. Lerman (Ed.), *Encyclopedia of mathematics education*. Dordrecht, The Netherdlands: Springer.
3. Sfard, A. (1994). Reification as the birth of metaphor. *For the Learning of Mathematics, 14*(1), 44–55.

Differentiation Rules and Attention to Scope 10

10.1 Statement of the Teaching Problem

In the 1988 movie "Stand and Deliver," mathematics teacher Jaime Escalante helps his students understand negative numbers by appealing to an analogy about digging holes at the beach:

> The sand that comes out of the hole, that's a positive number; the hole, that's a negative number.

Relating to his California students' oceanside experiences, Escalante created a conceptually productive way for them to expand their comprehension of how the integers behave. Like all analogies, it is limited in scope. It doesn't explain everything about how the integers work, but it does provide a tangible piece of intuition on which his students could build a more robust understanding.

How a teacher explains and communicates ideas is fundamental to the educational process. Effective teaching requires finding the right kinds of descriptions, the richest analogies, and the most accessible justifications. We have already discussed facets of classroom communication with respect to logic in Chap. 5, implicit assumptions in Chap. 7, and prioritizing conceptual understanding in Chap. 9. In this chapter, we look at mathematical explanations as a part of everyday communication in the classroom with a focus on finding the proper balance between offering useful insights and the obligation to be accurate and rigorous. What degree of imprecision is acceptable in the effort to provide explanations for students that have generative power for making sense of new ideas?

Consider the following pedagogical situation:

© The Author(s), under exclusive license to Springer Nature Switzerland AG 2022   143
N. H. Wasserman et al., *Understanding Analysis and its Connections to Secondary Mathematics Teaching*, Springer Texts in Education,
https://doi.org/10.1007/978-3-030-89198-5_10

A mathematics supervisor, Ms. Gunnarsson, observes several mathematics classes at a secondary school taught by different instructors. Planning to have a follow-up conversation about pedagogical quality, she records a list of explanations she heard teachers give as part of their instruction:

1. (*Exponents*) Exponents are just repeated multiplication.
2. (*Perimeter*) The perimeter of a shape is the sum of all the side lengths.
3. (*Probability*) Probability is the ratio of the number of desirable outcomes to the total number of possible outcomes.
4. (*Derivatives*) To take the derivative of a function with an exponent, bring down the exponent to the front and subtract one from the exponent.

Each of these statements represents an attempt to summarize and, in some sense, simplify a fundamental mathematical idea. The statements and explanations are perhaps not as "real-world" as the sand analogy, but they each allude to fundamentally important ideas. Repeated multiplication is a good foundation for exponents; recognizing perimeter as the composite measure of various side lengths is essential; the ratio of desirable to possible is a reasonable way to understand probability; determining derivatives of functions with numerical exponents is one of the first tasks in calculus. The teacher's job is to make ideas comprehensible and one way to achieve that is by adopting the more informal tone that characterizes these statements. But as we have discussed in earlier chapters, summary statements, analogies, and informal statements inevitably come with a degree of imprecision— they are imperfect snapshots of the idea as a whole. That is to say, there are constraints or limitations to statements like the ones Ms. Gunnarsson recorded (TP.1).

Before moving on, think about what you might discuss with the teachers Ms. Gunnarsson observed regarding the pedagogical quality of their statements.

10.2 Connecting to Secondary Mathematics

10.2.1 Problematizing Teaching and the Pedagogical Situation

One form of imprecision that creeps into statements like these comes from the terminology used. Is the phrase "repeated multiplication" in the exponent statement sufficiently clear? How might one determine which outcomes are "desirable" in the probability statement? Does "bring down the exponent to the front" inherently suggest that this exponent then multiplies the overall expression? We can categorize these kinds of concerns as being about the quality of the *explanation* of the mathematics. But evaluating such communicative traits of the explanation are not the only way to consider pedagogical quality. What about the *mathematics* of the explanation? We maintain that this is another pedagogically rewarding angle to pursue and, in this regard, closer inspection of each statement is warranted.

Just as important as figuring out what ideas to emphasize in an explanation is understanding what particular emphases can be left out. What degree of precision has been lost in these statements in the process of summarizing and simplifying? To what extent does each statement hold in different settings? What has been left unaddressed mathematically? We call this idea *attention to scope*:

Attention to Scope Attention to scope refers to the process of evaluating mathematical statements, descriptions, explanations, justifications, and proofs by explicitly identifying the "domain"—i.e., the set of objects or conditions—on which they are mathematically valid.

Attention to scope requires us to determine the assumptions or circumstances under which a statement is true and, conversely, when it might be false. Scrutinizing mathematical statements in a way that probes the limits of their validity is directly related to TP.1 and an essential component to monitoring pedagogical quality.

Before reading on, ask yourself under what conditions each of the four statements is valid. Are there examples or contexts in which each statement is invalid?

10.2.2 Attention to Scope

So that you can compare your results to ours, we consider each of the four statements separately, looking specifically at the scope to which they are applicable.

The exponent statement captures the fundamental idea that 2^n is the repeated multiplication of 2 by itself n times. This interpretation, however, is only meaningful when n is a positive integer. Without the ability to think about exponents in other ways, secondary students will struggle to interpret expressions like 2^0, 2^{-3}, and $2^{1/3}$. The number 2, repeatedly multiplied '0' times, would be, what—0? And what does 2 repeatedly multiplied '−3' times mean? These limitations reveal that the scope of this particular explanation is restricted to the natural numbers. Outside of this domain it does not offer much insight. Attention to mathematical scope might cue a teacher to recognize the need for a broader summary of exponentiation that is applicable to these other contexts.

The perimeter statement is a bit more robust in that it accurately summarizes the concept of perimeter for polygons of all shapes and sizes. In terms of limitations, it implicitly assumes that the shape in question is bounded by a finite number of straight edges. It doesn't make sense for shapes with curved edges like circles and ellipses, for instance; there are also extremely jagged mathematical shapes such as fractals with boundaries for which there is no simple way to calculate a well-defined length. In fact, we see now that the idea of what constitutes the boundary of a shape will require more mathematical attention before we can expand the scope of this statement. This sort of reasoning applies a form of logical "pressure" to the statement to see what cracks get exposed and what examples might serve as boundary cases (see TP.2).

The characterization of probability as the ratio of desirable outcomes to the total outcomes works well for calculating the probability of rolling a 5 with a fair six-sided die. With six possible outcomes, $\{1, 2, 3, 4, 5, 6\}$, and exactly one outcome of 5, the probability is $\frac{1}{6}$. But things get more complicated if we try to compute the probability of rolling a total of 5 with two dice. The answer is *not* $\frac{1}{11}$, despite the fact that 5 is one of the eleven possible outcomes $\{2, 3, 4, 5, 6, 7, 8, 9, 10, 11, 12\}$. (The actual probability is $\frac{4}{36}$.) The hiccup here is that outcomes on this list occur with differing frequencies; rolling a 7 is more likely than rolling a 2 or 12 for instance. For the probability statement to be applicable we must be in a situation where all the outcomes are equally-likely and the total number of outcomes is finite.

The derivative statement is a description of the syntax associated with the so-called power rule for differentiating functions of the form $f(x) = x^r$; for example, $\left(x^4\right)' = 4x^3$ and $\left(2x^3\right)' = 3 \cdot 2x^2 = 6x^2$. Among other benefits, this rule leads to a simple algorithm for differentiating an arbitrary polynomial. But how about $f(x) = e^x$? Applying the rule as stated suggests $f'(x) = x \cdot e^{x-1}$, which is *not* true. So what is the scope of the power rule? That's an interesting question. The previous example shows it does not apply when the variable x is in the exponent. Returning to functions of the form $f(x) = x^r$, we ask whether r must be a natural number? Can r be a negative integer? A rational number? An arbitrary real number?

Postponing the answer for a moment, let's pause to take stock of the central theme of the teaching scenario. We have seen that all four statements capture some core truth about a mathematical idea, but all four become inaccurate or completely false if they are interpreted too broadly. To be clear, we are *not* saying these statements are inherently bad or problematic. Teaching at all levels requires the instructor to provide succinct descriptions that capture the essence of an idea and convey it in an efficient way that students can quickly recall. The challenge is to be aware of the limitations of the summaries, analogies, explanations, and descriptions that are being offered.

Giving proper attention to scope can inform the degree to which ideas that are generative for making sense of new mathematics have been captured, and real analysis is an ideal place to practice this endeavor.

10.3   Connecting to Real Analysis

A real analysis course typically follows a "definition-theorem-proof" model. Definitions give precise meanings to new ideas, theorems articulate relationships between these ideas, and proofs provide a deductive argument to substantiate the statements in the theorems.[1] Acknowledging that this overview of a real analysis course itself falls into the oversimplification trap in the name of pedagogical clarity (read Abbott's [1] text and find passages that don't fit this summary!), we shall nevertheless adopt this mode of discourse for a moment to illustrate how attention to scope is paramount in the language of analysis. As a case study, we consider the power rule for differentiating $f(x) = x^r$.

10.3.1 Scope in Theorems and Proofs

The power rule states that if $f(x) = x^r$ then $f'(x) = rx^{r-1}$. Here is the proof Abbott gives [1, p. 148]:

Proof Let c be an arbitrary point in $\mathbb{R}$. The algebraic identity

$$x^r - c^r = (x - c)(x^{r-1} + cx^{r-2} + c^2 x^{r-3} + \ldots + c^{r-1})$$

allows us to calculate the desired formula

$$f'(c) = \lim_{x \to c} \frac{x^r - c^r}{x - c} = \lim_{x \to c} (x^{r-1} + cx^{r-2} + c^2 x^{r-3} + \ldots + c^{r-1})$$
$$= c^{r-1} + c^{r-1} + \ldots + c^{r-1}$$
$$= rc^{r-1}.$$

$\square$

Every proof—even a short one like this—requires effort to digest. Take some time to consider the details of this argument and, as you do, consider the following questions about the scope of the proof:

- Does the proof justify the formula $\left(x^5\right)' = 5x^4$?
- How about $\left(x^{-3}\right)' = -3x^{-4}$?
- For what specific values of r is this a valid proof?
- Where does the proof go awry for a value of r not in this set?

We have deliberately omitted the information about the restriction on r from the proof so that you can take a moment to ferret this information out for yourself before reading on.

---

[1] While it is true that the proofs demonstrate that the theorem is true, most of the proofs are probably intended to teach you more than that. Your professor is probably attempting to teach you techniques for proof-writing as well as ways of thinking.

The first clue that r must be a natural number is the algebraic identity at the outset of the proof. Summing the r expressions of the form c^{r-1} at the end is another clue. These steps only make sense if $r \in \mathbb{N}$, and this indeed is the proper scope of this particular proof. Not only statements but even proofs can have a domain for which they are valid.

Does that mean the power rule only holds for natural numbers? No, but it does mean we must construct a new proof if we want to justify the theorem for more general cases. Working through the sequence of arguments for the power rule that progressively extend the values of r from $\mathbb{N}$ to $\mathbb{Z}$ to $\mathbb{Q}$ to $\mathbb{R}$ is an especially effective way to appreciate the care and precision with which real analysis pays attention to scope. There is no room for ambiguity in this process. At each step, the scope is clearly stated and the successive proofs carefully rely on this hypothesis together with previously established results.

10.3.2 Extending the Power Rule

As a brief reminder of where we are, we have the power rule, which states:

$$\text{If } f(x) = x^r \text{ then } f'(x) = rx^{r-1}.$$

And at this point we have a proof for this statement in the case where $r \in \mathbb{N}$.

Claim The power rule holds if r is an integer.

Proof Let $f(x) = x^r$ with $r \in \mathbb{Z}$ and consider the case when $r < 0$. We can write $f(x) = x^r = \frac{1}{x^{-r}}$, and observe that $-r$ is a natural number. Using our previously proved version of the power rule to differentiate x^{-r}, the quotient rule for derivatives (see Abbott's Theorem 5.2.4) implies

$$f'(x) = \frac{(x^{-r})(0) - (1)\left(-rx^{-r-1}\right)}{\left(x^{-r}\right)^2} = rx^{-r-1+2r} = rx^{r-1}.$$

Note that for the case $r = 0$, $f(x) = 1$ and so $f'(x) = 0$ which agrees with the the power rule formula.[2]   □

Claim The power rule holds if r is a unit fraction.

Proof Let $f(x) = x^r$ where $r = \frac{1}{n}$ for $n \in \mathbb{N}$. This means f is the inverse function of $h(x) = x^n$ (for $x > 0$). Applying the power rule for $n \in \mathbb{N}$ together with the

---

[2] When $x = 0$ we get the indeterminant form $f(0) = 0^0$ which we can exclude from the domain (as we do when $r < 0$) or define to be 1.

formula for the derivative of an inverse function (see Abbott's Exercise 5.2.12), gives

$$f'(x) = \frac{1}{h'(f(x))} = \frac{1}{n(x^{\frac{1}{n}})^{n-1}} = \frac{1}{n}x^{\frac{1}{n}-1} = rx^{r-1}.$$

$\square$

Claim The power rule holds if r is a rational number.

Proof Let $f(x) = x^r$ with $r \in \mathbb{Q}$. This means we can write $r = \frac{m}{n}$ for $m \in \mathbb{Z}$ and $n \in \mathbb{N}$, and so $f(x) = (x^m)^{1/n}$. Combining the power rule for the unit fraction $\frac{1}{n}$, the power rule for the integer m, together with the chain rule (see Abbott's Theorem 5.2.5) gives

$$f'(x) = \frac{1}{n}(x^m)^{\frac{1}{n}-1} \cdot mx^{m-1} = \frac{m}{n}x^{\frac{m}{n}-m+m-1} = \frac{m}{n}x^{\frac{m}{n}-1} = rx^{r-1}.$$

$\square$

Claim The power rule holds if r is a real number.

Proof This proof requires more sophisticated machinery than the others, and so we content ourselves with a sketch of the the argument.

Let $f(x) = x^r$ with $r \in \mathbb{R}$. By the density of $\mathbb{Q}$ in $\mathbb{R}$, there exists a sequence (a_n) of rational numbers that converges to r. For each n, define the function $g_n(x) = x^{a_n}$. Because $(a_n) \to r$, it follows that $g_n(x) \to f(x)$, and it should also seem reasonable (though maybe not obvious) that $g'_n(x) \to f'(x)$. (See Abbott's Theorem 6.3.1.) Now let's compute the limit of the sequence $g'_n(x)$ in a different way. Applying the power rule for rational exponents to $g_n(x)$ allows us to conclude $g'_n(x) = a_n x^{a_n-1}$, and remembering that $(a_n) \to r$ it follows that $g'_n(x) \to rx^{r-1}$. Having already convinced ourselves that the limit of $g'_n(x)$ should be $f'(x)$, we conclude that $f'(x) = rx^{r-1}$, establishing the power rule in this most general case.

$\square$

Understanding the details of these arguments is not the main point of this discussion. What's important is appreciating the care and precision that goes into expanding the scope of the power rule in this hierarchical fashion, each step becoming a stepping stone for the ones that come next. The meticulous attention to detail and logically developed propositions that build in complexity is characteristic of real analysis and represents an idealized way to view mathematics more generally. Some would say this is what mathematics is. This also suggests a heuristic for navigating questions of scope in the domain of pure mathematics: in a perfect mathematical world, *be stringently precise with statements and completely rigorous with proofs.*

This, of course, only applies in a so-called perfect mathematical world, wherever that might be. There is a time and place for this approach to mathematics, but teachers must consider the needs of their learners. Even in your real analysis class, your professor likely uses informal statements to convey ideas or help you appreciate nuances that might be obscured by too much formality. Although it is good practice to always be aware of issues of scope in the explanations we provide, being an effective teacher requires finding a healthy balance between precision and practicality.

10.4 Connecting to Secondary Teaching

With the exacting standards of real analysis ringing in our ears, it is time to turn the critical lens onto what we say and write in the classroom, as well as onto what our students say and write. Attending to scope is related to TP.1 in the way it forces us to acknowledge and revisit assumptions, but this process needs to be moderated with regard for the needs of the students in the classroom.

10.4.1 Attention to Scope as a Pedagogical Practice

All four statements in the initial teaching situation have a limited scope. Each one is valid in the proper context but loses its validity when that context, or domain, gets too broad. One option in teaching would be to specify this scope as a condition in the statement—to be stringently precise. For example, the Perimeter statement might become "The perimeter of a polygon is the sum of all the side lengths"; the Exponents statement, "Natural-numbered exponents can be understood as just repeated multiplication"; and so on. While this is an option, and these statements are accurate, the particular conditions and their scope would need to be explicitly unpacked and discussed. Otherwise, students might not understand the need for the clarification in the first place, or even remember it.

We now consider two alternate ways of attending to scope in teaching.

Approach 1 Using attention to scope to generalize.

One way to address the situation is to refashion the statements to be more *general* so that they hold up as the scope gets bigger. In the case of the perimeter statement, for instance, we might try the following alternative:

> (*Modified perimeter statement*) The perimeter of a shape is the length of the boundary that encloses the shape.

This definition is more general. It can be applied to a polygon or an ellipse or a shape bounded between two parabolas.[3] One implication of this generalizing approach is that specific instances can be discussed as particular applications. So, starting from this more general definition, a teacher could then ask students how they would determine the perimeter of a polygon. It would not be hard for a student to translate this general definition to the particular conclusion that the perimeter of a polygon is the sum of the lengths of its edges. Moreover, when it came time to consider circles, this general statement would provide a way to think about how to compute its circumference in a way that "summing side lengths" does not.

This more general description of perimeter has the desirable attribute of being true in a broader range of cases, and of allowing the learning of similar concepts to be meaningfully connected, but there is also a cost attached. Namely, in addition to the general statement, a second step is almost always needed. It was only after giving the general description that the useful procedure of summing side lengths to find the perimeter of a polygon could be extracted. Moving from the general case to the particular case requires attending to both.

Approach 2 Using attention to scope to sharpen and foreshadow.

Instead of broadening their scope, an alternative would be to *sharpen the scope* of statements by *explicitly foreshadowing the limitations* that will ensue. This might make sense when a more general statement, or the particular assumptions, are not something students would be able to make sense of. For example, in the case of the Exponents statement, a proper clarification that narrows its scope would be "Exponents that are natural numbers can be understood as repeated multiplication." This improvement cures the statement of its earlier imprecision, but students navigating in unfamiliar mathematical terrain might not recognize the non-trivial nature of the assumption about $\mathbb{N}$ being made explicit. As an alternative, we might keep the original statement in tact but explicitly clarify how it is limited in scope:

> (*Modified exponents statement*) Exponents are repeated multiplication. This is a good place to start—it makes sense when we have exponents like 2, 3, 4, 5, etc.—but we will have to revisit this later because the description won't make sense when we consider exponents that aren't natural numbers.

---

[3] This is true provided we have a suitable definition for the length of the bounding curves.

This statement points out that the current description will lose its validity at some point in the future. It endorses the explanation as sufficient for the current context but indicates that it will eventually need to be revised. Depending on whether students have been introduced to negative numbers, it could make sense for the teacher to inject '3^{-2}' immediately into the conversation as a case that falls outside the domain where "repeated multiplication" is a useful description of exponentiation. Alternatively, the teacher could directly pose the question of scope to students, asking them to think about how broadly this description holds up. A takeaway of TP.1 is that the onus to make these limitations explicit is on the teacher, regardless of whether it is ultimately the teacher or a student who points them out. When a limited statement is made, it is the teacher who must be responsible for ensuring that such limitations are intentionally discussed, revisited, and illustrated. The timing of when to engage in this type of scrutiny is an open question. It could be better to give the students sufficient time to internalize what the statement does say before exploring what it doesn't.

10.4.2 Considering the Scope in Proofs Is Useful Too

To preview a topic discussed in Chap. 12, consider the formula for the area of parallelogram $A_{par} = b \cdot h$. One way to derive this formula is to cut a triangle off one end of the parallelogram and reassemble it on the other, thereby converting the parallelogram into a rectangle with dimensions $b \times h$.

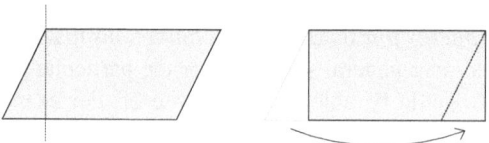

This is a richly effective justification for the parallelogram's area formula. The problem is that the procedure does not always produce a rectangle. The scope for this justification is limited—it doesn't work in the case of a particularly tall and skewed parallelogram like the one below.

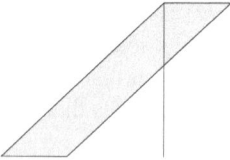

Both approaches—generalizing or sharpening and foreshadowing—could be appropriate. As one option, a more general justification might be sought. In fact, as we will see in Chap. 12, there are other justifications for a parallelogram's area that have a larger scope. But we won't spoil it just yet! The point is that if a more general justification is given, the "cut-reassemble" argument could be discussed as a secondary one. As another option, a teacher could give this argument but then explicitly sharpen it by foreshadowing its limitations, perhaps qualifying it as "a good place to start." In either case, these two approaches offer some ways to think about what to do when justifications, not only statements, have a limited scope.

Although our examples in this chapter have been in terms of a *teacher's* statements, the same ideas apply to *students'* contributions. If a student makes a mathematical contribution that has limitations, the teacher should attend to the scope of those statements with the same sense of balance we have been discussing. This means not always correcting students' slight mis-statements in the moment but perhaps revisiting them later when the student is ready to consider more nuanced issues.

It is a challenge to constantly stay alert to the scope of mathematical statements. Doing so requires teachers to think carefully about the concepts they are teaching as well as the concepts their students will encounter in the future. For an early elementary teacher, the idea that "multiplication makes bigger" is valid and helpful, but this notion does not hold up when multiplication expands beyond the natural numbers in later years. The same goes for high school teachers. Some ideas in Euclidean geometry translate quite nicely to non-Euclidean geometries, but others do not. Attending to scope means thinking through these implications so that students are prepared for future mathematical developments.

Problems

10.1 At the beginning of this chapter, we mentioned Jaime Escalante's description of negative and positive numbers in terms of digging holes at the beach. Describe in what ways this analogy is useful for understanding the integers. Then describe any ways that the analogy has a limitation (i.e., is limited in scope).

10.2 Consider the following statement made by an elementary teacher: "Remember, to multiply a number by ten, just add a 0 to the end." i) Determine for what set(s) of objects the statement is true. If there are any, provide an example of a set(s) of objects for which the statement is not true. ii) Describe what difficulties, if any, students who only understand multiplication by ten in terms of the statement above might run into later in their mathematical careers? How might you alter the statement, if at all?

10.3 Similar to the Exponents statement, it is common in mathematics to claim that "multiplication is repeated addition." First, explain the scope for which this description is a valid statement. Next, think about how you might explain multiplication to students. Give an example of a more *general* description of multiplication that addresses some of the statement's limited scope. Make sure to describe how repeated addition might fit as a particular example of the more general description.

10.4 Consider the following informal explanation for why $\left(a^b\right)^c = a^{bc}$:

> Let's illustrate this with a particular example: $\left(2^3\right)^5$. We will use the fact that $(a^x) \cdot (a^y) = a^{x+y}$. We can write $\left(2^3\right)^5$ as five copies of $\left(2^3\right)$, so that:
>
> $$\left(2^3\right)^5 = \left(2^3\right) \cdot \left(2^3\right) \cdot \left(2^3\right) \cdot \left(2^3\right) \cdot \left(2^3\right) = \left(2^{3+3+3+3+3}\right) = 2^{3 \cdot 5}$$

What number sets of a would be appropriate to include here as part of the scope of this explanation? What number sets for b? What number sets for c?

10.5 Suppose you observe a high-school teacher solving the algebraic equation $7x - 4 = 15$ in class. The teacher explains to the students:

> The first thing we do to solve is to add 4 to both sides of the equation to obtain $7x = 19$. Remember an equation is an equality of two expressions. So if the expressions are to remain equal at each step of the solution process, whatever you do on one side of the equation, you must do on the other as well. Now, when we are solving an equation we are always *allowed* to perform the same operation on both sides of the equation, but depending on the equation some operations are *more useful* than others for finding the solutions for x. In this case, a useful second step would be to divide by 7 on both sides of $7x = 19$.

Evaluate the pedagogical quality of the teacher's explanation. Give at least one thing that is good about the teacher's explanation and one that could be improved. Provide a detailed rationale for your evaluation of this explanation.

10.6 A common description of whether a graph is a function is in terms of the vertical line test (VLT). Specifically, the VLT statement is: "A graph represents a function if each vertical line intersects the function no more than once." Many mathematics educators dislike teaching students the vertical line test. In part, this is due to the fact that students are frequently given the rule without conceptual reasons. However, this rule is also inaccurate because it is limited in scope. For what types of graphs would this rule be accurate? What future mathematics might be confusing for students who learn the vertical line test? [Note: as a recall from Chap. 7, a function is a set of ordered pairs where each element in a domain set A is associated with a unique element in a codomain set B. You might think about various types of functions!]

10.7 The area of a trapezoid is found by: $A_{trap} = \frac{1}{2}h(b_1 + b_2)$. First, provide a justification for this area formula—give a visual depiction indicating *why* it is true. Then, think about various trapezoids, and determine whether your justification method would be valid in those cases—or whether it would have a limited scope. Describe any limitations. Last, the area formula itself would be valid for all trapezoids using the exclusive definition of trapezoid (i.e., exactly one pair of parallel sides). Describe whether or not it would also be valid with the inclusive definition of trapezoid (i.e., at least one pair of parallel sides).

10.8 With the Probability statement (probability is the ratio of the number of desirable outcomes to the total number of possible outcomes), not only do students learn to find probability by the ratio of desirable over total possible outcomes, but they, in fact, mis-apply this in situations when it doesn't make sense. For example, assuming that having a boy and a girl are equally-likely, what is the likelihood that a family with two kids will have both of the same gender? Most students think of the following as the total possible outcomes: two boys, two girls, one of each. Hence, they get an answer of P(two kids same gender) $= 2/3$. Here, they have not mis-applied the ratio, but rather they have presumed the equally-likely assumption that was implicit in the statement for it to be valid. This is known as the equi-probability bias; that when we think of outcomes, we tend to think of each as equally-likely. In this case, that is incorrect. The following are the probabilities for the three cases: P(two boys) $= 1/4$, P(two girls) $= 1/4$, P(one of each) $= 1/2$. Often in probability problems, when outcomes are not equally-likely, there is a *way* to think about the outcomes as being equally-likely. Describe a way to think about the total possible outcomes to this situation, and i) justify that each outcome is equally-likely, and ii) use these to justify the correct probabilities (given above).

10.9 Consider the progression of exponents in secondary mathematics. First, provide a brief explanation for how you understand the exponential expression, a^b, where a and b are numbers in the following sets: i) $\mathbb{N}^{\mathbb{N}}$; ii) $\mathbb{N}^{\mathbb{Z}}$; iii) $\mathbb{N}^{\mathbb{Q}}$; iv) $\mathbb{N}^{\mathbb{R}}$; and v) $\mathbb{Z}^{\mathbb{Q}}$. Second, if you try to think about $(-8)^{1/3}$ as $\sqrt[3]{(-8)^1}$, and $(-8)^{2/6}$ as $\sqrt[6]{(-8)^2}$, the result is that $(-8)^{1/3} \neq (-8)^{2/6}$, despite the fact that $1/3 = 2/6$. Discuss how this might influence your thinking about $\mathbb{Z}^{\mathbb{Q}}$. Third, as a teacher, discuss some things that you would do to help students understand the meaning of a rational exponent.

10.10 In Chap. 6 we discussed definitions. Revisit Problem 6.7 which discusses two different definitions for the absolute value of a number, $|x|$. Describe the scope of each of these definitions—i.e., for which numbers each makes sense. Describe which is the more general definition, and how the other could be described as an application of the more general definition to a more specific situation. Lastly, extend the more general definition to determine the absolute value of a point $q = (x, y, z)$ on a three-dimensional graph and its algebraic expression.

10.11 In Chap. 8 we discussed solving trigonometric equations. Composing inverse functions is generally a way to "undo" aspects of an equation. In particular, we might simplify 'arcsin $(\sin (x))$' to be 'x'. In doing so we are claiming

$$\arcsin (\sin (x)) = x.$$

What is the scope of this identity? Is it all real numbers? Can you explain why the scope is what it is? (Hint: you might consider graphing it.) Does the identity $\sin(\arcsin(x)) = x$ have the same scope, or a different one?

10.12 Immediately prior to stating Theorem 5.2.5 (Chain Rule), Abbott provides some commentary about the Chain Rule. Namely, he states [1, p. 150]:

$$(g \circ f)'(c) = \lim_{x \to c} \frac{g(f(x)) - g(f(c))}{x - c} = \lim_{x \to c} \frac{g(f(x)) - g(f(c))}{f(x) - f(c)} \cdot \frac{f(x) - f(c)}{x - c} = g'f(c) \cdot f'(c)$$

With a little polish, this string of expressions could qualify as a proof except for the pesky fact that the $f(x) - f(c)$ expression causes problems in the denominator if $f(x) = f(c)$ for x values in arbitrarily small neighborhoods of c...

What teaching principle do you believe Abbott's commentary is illustrating here? Elaborate.

Turning the Tables

Reflecting on *teaching* from your *learning* in real analysis: TP.5

As another sojourn, we consider TP.5—avoid giving rules without accompanying mathematical explanations—in relation to some of your learning in real analysis. In discussing TP.5 in prior chapters, we have primarily focused on ensuring that explanations provide meaningful conceptual justification rather than a rule to follow. The discussion here is not meant to replace those ideas but to offer an additional reflection about the entire arc of an explanation.

Let's consider the development in Abbott's book around the relationships between differentiability, continuity, and the intermediate value property. When introducing the Intermediate Value Theorem (IVT), Abbott goes out of his way to introduce the Intermediate Value Property (IVP): "A function f has the *intermediate value property* on an interval $[a, b]$ if for all $x < y$ in $[a, b]$ and all L between $f(x)$ and $f(y)$, it is always possible to find a point $c \in (x, y)$ where $f(c) = L$" (Definition 4.5.3). Essentially, it means the IVT can be restated as: "Functions continuous on an interval $[a, b]$ have the intermediate value property." (Check that this is what you found for Problem 7.8.) Discussing the IVP at this point is unnecessary, but Abbott is using the opportunity to foreshadow things to come. He is hinting that continuity and the IVP—intertwined in the IVT—should, for some unnamed future reason, be understood as separate.

When he gets to the derivative function (Sect. 5.1), Abbott poses questions about the relationship between differentiation and continuity. The queries lead the reader to recognize that derivatives need not be continuous, yet they also seem to maintain a property which Abbott conjectures to be the IVP. In doing so, the reader is drawn into considering mathematical relationships, not just rules. The justification for the conjecture comes subsequently in two stages— The Interior Extremum Theorem (Theorem 5.2.6) and Darboux's Theorem (Theorem 5.2.7). The former is the simpler result, analogous to the Intermediate Zero Theorem from Chap. 7 and directly connected to optimization. The latter is deeper, its proof relying on the former results and asserting that the IVP applies to the derivative function. In the justification process Abbott engages the students, leaving it to them to complete the argument in an exercise. In doing so, Abbott reveals that the derivative's use in optimization problems carries with it the implication that derivative functions defined on an interval maintain the IVP—even though they may not be continuous!

Of course, this was the reason for introducing the IVP earlier; to characterize a property of derivative functions that explains their use in solving optimization problems. From this example, we point out that meaningful mathematical explanations can have a relatively broad arc—where important ideas are seeded earlier, queried and conjectured later, and then ultimately justified.

References

1. Abbott, S. (2015). *Understanding analysis* (2nd ed.). New York, NY: Springer.

Taylor Polynomials and Modeling the Complex with the Simple

11

11.1   Statement of the Teaching Problem

Novelty can bring with it feelings of uncertainty. What do we do when confronted with situations for which our previous experiences and knowledge are insufficient? "Build the plane as you fly it"—or "make it up as you go along"—is a common response that captures the inclination to innovate solutions in real time. Ad hoc solutions, though, tend to be unsystematic. A more structured approach, reminiscent of Polya, could use problem-solving heuristics to address the novelty.[1] But no matter which way you slice it, creativity and ingenuity are required.

New situations arise regularly, both in mathematics and in teaching. Consider the transition from working with linear objects like straight lines and polygons to curved ones like parabolas and circles. Concepts such as slope, length, and area, which are straightforward when everything is linear take on an air of uncertainty when the straight lines are replaced by curves. What do we really mean by the slope of a curve? What do we mean by the length of a curve and how do we measure it? We typically measure length by decreeing a standard measuring stick—with straight length units like a foot, a meter, a mile—but it's hard to use a stick to measure the length of a curve. Sticks don't bend well.

The first curved object students usually encounter is a circle, and even this simple example can be disorienting. Where do the familiar formulas $C = \pi d$ and $A = \pi r^2$ come from, and what is the proper definition of π? How do we determine the value of this enigmatic number?

Consider the following pedagogical situation:

---

[1] Polya [2] identified four problem-solving heuristics: (1) understand the problem; (2) devise a plan to solve it; (3) carry out the plan; and (4) look back.

© The Author(s), under exclusive license to Springer Nature Switzerland AG 2022   159
N. H. Wasserman et al., *Understanding Analysis and its Connections to Secondary Mathematics Teaching*, Springer Texts in Education,
https://doi.org/10.1007/978-3-030-89198-5_11

Ms. Liu sets up an activity to help her students learn how to determine the circumference of a circle. Three groups of students are each given a ruler and a different sized plate. Ms. Liu then asks them to measure the diameter and circumference of their plate. To measure the circumference, one group rolls the plate along the ground and keeps track of the length. Another group attempts to roll the ruler around the outside of the plate. The third group wraps a shoelace around the plate and then stretches it out to measure it. They get the following results:

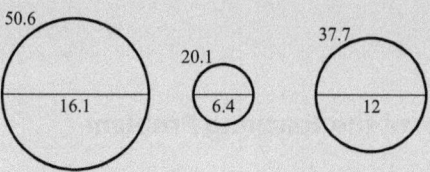

Ms. Liu then asks the students to compute the ratio of the circumference and the diameter in each case. The resuts are $\frac{50.6}{16.1} \approx 3.1428$, $\frac{20.1}{6.4} \approx 3.1406$, and $\frac{37.7}{12} \approx 3.1416$. The students remark on the fact that the ratios are all about the same. Ms. Liu confirms that the ratio of circumference to diameter is the same for any circle, about 3.14, and that this constant value is known as π.

This hands-on activity provides an effective entry point into the geometry of circles. The tangible objects make the ideas less abstract, and it emphasizes that measuring the circumference is more complex than measuring the diameter. The activity also allows the students to discover for themselves that the ratio $\frac{C}{d}$ is a constant and, although measurements can be imprecise, these examples give a reasonably good estimate of π. Yet, why the measurement around a circle is a little more than three times the diameter is still unclear. And, perhaps more importantly, how might we approach the challenge of measuring curved lengths based on what we already know about straight lengths? Mathematics and mathematics teaching are nice in the sense that there are some normative approaches that provide structure for how we might address these challenges in teaching.

Before moving on, think about what questions you would ask the students at this point and what you might do next to follow up on this activity.

11.2 Connecting to Secondary Mathematics

11.2.1 Problematizing Teaching and the Pedagogical Situation

A good place to start is with the observation that $\frac{C}{d}$ is a constant. This is the conclusion the students' empirical evidence seems to support, and it makes intuitive sense when we think about how ratios work in shapes that are similar. If we double the radius, it seems like we double the circumference. But is this obvious? Does it need a proof? What might such a proof consist of?

Having made the case that $\frac{C}{d}$ is constant for all circles, Ms. Liu could rewrite the identity $\pi = \frac{C}{d}$ in the more familiar form $C = \pi d$ and point out to her students that this enables them to find the circumference of any circle from its diameter (or radius). Or the other way around. Given that the earth's circumference is 24, 902 miles, she could ask them to find its radius. A more hands-on follow-up activity might be to have students measure the circumferences of the plates again, but this time using a measuring tape with a non-standard unit—more specifically, letting the diameter of each circle be the unit of measure. There are also dynamic technologies, such as GeoGebra, which could be used to demonstrate that, even when changing the diameter of a circle, the ratio of diameter to circumference remains constant. But these approaches use the circle as the starting point.

Two open questions such an approach leaves unattended are: What is a reasonable way to measure the length of a curve? and What is the actual value for π? The students all got ratios around 3.14, but how reliable is this estimate? These interesting questions, together with the inquiry about how to prove $\frac{C}{d}$ is constant in the first place, can all be addressed using an approach we describe as *modeling the complex with the simple*. This approach uses known objects and relationships as a way to model new and more complex objects in order to investigate them. In the present example, our complex object is the circle and the known objects will be constructed from straight line segments. Before moving on, try to think of a natural way to approximate a circle using objects made from linear components.

11.2.2 Bounding and Approximating π with Regular n-gons

Our strategy is to use regular polygons to model the circle. So instead of starting with the circle, we begin with these other familiar shapes. Let's start with a hexagon—two in fact, one inscribed and one circumscribed about a circle of radius r.

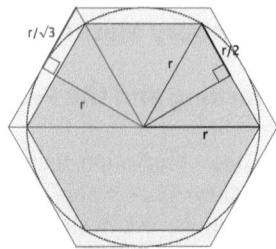

Measuring circumference is a new endeavor, but computing the perimeter of a hexagon is familiar territory. For the inscribed hexagon the perimeter is easily seen to be $6r$ since the hexagon is composed of six equilateral triangles. The diagram also makes clear that this perimeter *underestimates* the circumference of the circle. Thus $6r < C$, which in terms of the diameter means $3d < C$.

Computing the perimeter of the circumscribed hexagon requires a bit more effort because r represents the *height* of one of the equilateral triangles that form the hexagon. Taking advantage of the friendly $30° - 60° - 90°$ triangles, the length of the side of one of these triangles is $\frac{2r}{\sqrt{3}}$. This means the perimeter of the circumscribed hexagon is

$$6\left(\frac{2r}{\sqrt{3}}\right) = \frac{12r}{\sqrt{3}} = \frac{6d}{\sqrt{3}} = 2\sqrt{3}d,$$

which must *overestimate* C. Notice how this approximation process allows us to deduce something about a more complex object (the circle) from a simpler one (the hexagon). This is a contrast to empirical measurements, where we may not be sure how much error an estimate contains. From this case, we know

$$3d < C < 2\sqrt{3}d.$$

Recalling that $C = \pi d$, we have now rigorously established that the value of π must be strictly between 3 and $2\sqrt{3} \approx 3.464$.

And there is no reason to stop here. Increasing the number of sides of the polygon reduces the error between the polygon's perimeter and the circumference of the circle. This was the technique used by Archimedes, the fabled Greek mathematician from the third century BCE, who estimated π with an error on the order of 0.0005 using a polygon with 96 sides. With the aid of some trigonometry (which Archimedes did not have), we can extend this computation to any arbitrary n-gon.

A regular n-gon consists of n isosceles triangles, each with angle $2\pi/n$ at the center of the circle. A sturdy but manageable trigonometry exercise shows the inscribed polygon has perimeter $2nr \sin\left(\frac{\pi}{n}\right)$, and the circumscribed polygon has perimeter $2nr \tan\left(\frac{\pi}{n}\right)$. (Try to verify these formulas!) Recasting these perimeters in terms of the diameter d gives

$$\left[n \sin\left(\frac{\pi}{n}\right)\right] d < C < \left[n \tan\left(\frac{\pi}{n}\right)\right] d.$$

The table below gives the decimal approximation for the perimeters, written as multiples of d, for the same n-gons that Archimedes computed. Presenting it in this form illuminates how the successive upper and lower bounds for π are getting closer together.

n	Perimeter (inscribed n-gon) $[n \sin(\pi/n)] \cdot d$	Perimeter (circumscribed n-gon) $[n \tan(\pi/n)] \cdot d$
4	$2.824d$	$4d$
6	$3d$	$3.4641d$
12	$3.1058d$	$3.2154d$
24	$3.1326d$	$3.1597d$
48	$3.1394d$	$3.1461d$
96	$3.1410d$	$3.1427d$

Implicit in these sharper and sharper estimates for π is a proper proof that $\frac{C}{d}$ is really constant. Rewriting the previous inequality in the form

$$n \sin\left(\frac{\pi}{n}\right) < \frac{C}{d} < n \tan\left(\frac{\pi}{n}\right),$$

and observing $\lim_{n \to \infty} n \sin(\pi/n) = \lim_{n \to \infty} n \tan(\pi/n)$ makes it clear that there is only one possible value for $\frac{C}{d}$. Another issue this approximating process resolves is the general question of identifying how length can be properly defined for curves. In place of using a stretched out shoestring to measure length (which only provides empirical estimates and not a deductive rationale), we can approximate our curve with a series of line segments whose lengths we can unambiguously sum. We then take a limit as the lengths of these intervals go to zero. If the limit exists, as it does for the circle, this limiting value is defined to be the length of the curve in question.

Defining length this way is a fairly advanced idea but exploring the circle with approximating n-gons offers a range of insights at all levels. For instance, a teacher could be content to inscribe a single hexagon and circumscribe a single square (Fig. 11.1). With no trigonometry and very little algebra students can deduce $6r = 3d < C$ and $C < 8r = 4d$, from which it follows that π is definitively between 3 and 4.

Students also get the experience of transitioning to a multiplicative referent for determining perimeter. Most students think about the perimeter of a hexagon as $s + s + s + s + s + s = 6s$ and a square as $s + s + s + s = 4s$. That is, they *sum* the individual side lengths that make up the perimeter. But having students determine the perimeter of the hexagon based on the diameter instead of a side asks them to change the referent. The expressions $3d$ and $4d$ are *multiplicative* comparisons about how the length of the perimeter compares to the length of the diameter. The

Fig. 11.1 (**a**) Lower-bound for circumference of $3d$ (or $6r$) and (**b**) upper-bound of $4d$ (or $8r$)

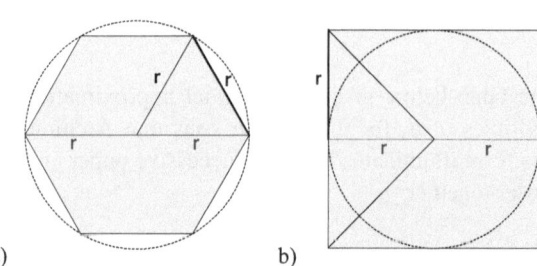

a)   b)

referent for perimeter can be a measurement other than a side length; indeed, as with diameter, it need not even be part of the perimeter!

All of these insights that arise from approximating the circle with polygons are examples of a general practice in mathematics:

Modeling the Complex with the Simple Modeling the complex with the simple is using simpler, familiar concepts as a way to approximate more complex, less-understood concepts and to deduce and develop ideas about them.

This practice might be related to another problem-solving heuristic: if you can't solve the current problem, solve a simpler one instead! In fact, we have seen this process in action on multiple occasions. In Chap. 3 we studied irrational numbers by approximating them with rational numbers. In Chap. 9 we discussed how the slope of a curve can be rigorously defined by approximating the curve with increasingly refined secant lines. Chapter 12 on integration is about computing areas of curved regions by approximating them with rectangles. This process makes ideas about more complex objects connected to, and informed by, existing mathematics. It is not an exaggeration to say that modeling the complex with the simple is a central theme of real analysis.

11.3   Connecting to Real Analysis

Setting aside the properties of circles for a moment, we turn out attention to the theory of Taylor polynomials which is another high-profile example of how modeling the complex with the simple expands the power of mathematics.

The standard normal distribution is central to statistics because its' symmetrical nature fits many natural phenomena. This distribution, affectionately known as the "bell curve," is given by the equation

$$\phi(x) = \frac{1}{\sqrt{2\pi}} e^{-x^2/2}.$$

Fig. 11.2 The standard normal distribution, or "bell curve"

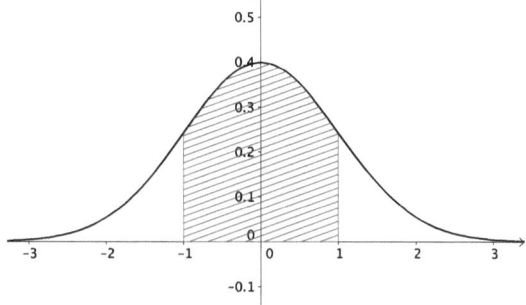

To compute a probability from a density function like $\phi(x)$, we find the area under the curve over a specified region. The probability that a standard normal random variable takes a value between -1 and 1, for instance, is given by the area of the shaded region in Fig. 11.2. Before moving on, think about how you might compute the specific area under the bell curve using techniques you learned in calculus.

11.3.1 Taylor Polynomials

The typical way to compute areas under curves is with an integral. To find the probability that our random variable takes a value between -1 and 1 we need to compute

$$\frac{1}{\sqrt{2\pi}} \int_{-1}^{1} e^{-x^2/2} dx.$$

There is just one problem—finding an anti-derivative for $e^{-x^2/2}$ is more challenging that it might look. In fact, it turns out that *there is no closed form for the anti-derivative of this function*! Faced with this reality, how might we compute the value of this integral another way? In the spirit of modeling the complex with the simple, our strategy is to approximate the more challenging function $e^{-x^2/2}$ with a series of simpler *polynomial* functions of nth degree, $p_n(x) = a_0 + a_1 x + a_2 x^2 + \ldots + a_n x^n$. In this case, what makes polynomials "simple" is the relative ease with which we can subject them to the processes of calculus. Computing derivatives and integrals is straightforward for polynomials, as is evaluating them. In the wide spectrum of functions used in mathematics, polynomials are among the most elementary.

So, how might we find a polynomial that approximates some function f? Among the many algorithms that exist, the oldest is the method of *Taylor coefficients*. In this method, we focus on the center point $x = 0$ and choose coefficients a_0, a_1, a_2, etc., so that $p_n(x)$ and $f(x)$ agree in y-value and their first n derivatives at $x = 0$.

This approach leads to Taylor's formula (see Abbott's [1] Theorem 6.6.2), which generalizes one way to determine coefficients of such a polynomial:

Taylor's Formula Supposing $f^{(n)}(x)$ represents the nth derivative of a function f, Taylor's formula asserts the coefficient of x^n in a polynomial that approximates the function f is given by $a_n = \frac{f^{(n)}(0)}{n!}$. So the nth degree polynomial would be:

$$p_n(x) = \frac{f(0)}{0!} + \frac{f^{(1)}(0)}{1!}x + \frac{f^{(2)}(0)}{2!}x^2 + \dots \frac{f^{(n)}(0)}{n!}x^n.$$

To warm up, let's find a second degree polynomial $p_2(x) = a_0 + a_1x + a_2x^2$ that approximates $f(x) = e^{-x^2/2}$, our "bell curve" function *without* the constant $\frac{1}{\sqrt{2\pi}}$, on the interval $[-1, 1]$. Using the chain rule, we compute the first two derivatives: $f'(x) = f^{(1)}(x) = -xe^{-x^2/2}$; $f''(x) = f^{(2)}(x) = -1e^{-x^2/2} + x^2e^{-x^2/2}$. Evaluating at $x = 0$ gives the Taylor formula coefficients:

$$e^{-x^2/2} \approx \frac{f(0)}{0!} + \frac{f^{(1)}(0)}{1!}x + \frac{f^{(2)}(0)}{2!}x^2$$

$$\approx \frac{e^{-0^2/2}}{0!} + \frac{-0e^{-0^2/2}}{1!}x + \frac{-1e^{-0^2/2} + 0^2e^{-0^2/2}}{2!}x^2$$

$$\approx 1 + 0x - \frac{1}{2!}x^2.$$

This quadratic turns out to be a reasonable stand-in for the "bell-curve" (minus the constant) on $[-1, 1]$—around $x = 0$. But why stop there? We could compute even more terms of the polynomial. The sixth degree Taylor polynomial produces:

$$e^{-x^2/2} \approx 1 + 0x - \frac{1}{2!}x^2 + 0x^3 + \frac{3}{4!}x^4 + 0x^5 - \frac{15}{6!}x^6$$

$$\approx 1 - \frac{1}{2}x^2 + \frac{1}{8}x^4 - \frac{1}{48}x^6.$$

Utilizing this polynomial approximation,[2] we revisit our previously incalculable integral and estimate it as

$$\frac{1}{\sqrt{2\pi}}\int_{-1}^{1} e^{-x^2/2}dx \approx \frac{1}{\sqrt{2\pi}}\int_{-1}^{1}\left(1 - \frac{1}{2}x^2 + \frac{1}{8}x^4 - \frac{1}{48}x^6\right)dx$$

---

[2] Another way to find this approximation would be to calculate the third degree Taylor polynomial for e^u, which is $1 + u + \frac{u^2}{2!} + \frac{u^3}{3!}$, and then substitute in $u = -x^2/2$.

$$\approx \frac{1}{\sqrt{2\pi}} \left(x - \frac{1}{6}x^3 + \frac{1}{40}x^5 - \frac{1}{336}x^7 \right)\Big|_{-1}^{1}$$

$$\approx 0.68247.$$

For those familiar with statistics, the area under the standard normal density from -1 to 1 is the probability of being within one standard deviation of the mean. The so-called "68-95-99.7" rule for normal distributions reminds us that the value of our integral should be about 0.68, which is what our results show! Approximating with our Taylor polynomial worked extremely well. In fact, it gives an estimate of the actual value of the integral with an error of about 0.0002.

Similar to the strategy of approximating a circle with regular polygons, it appears as though we can improve our estimate of this integral by using even higher degree polynomials. But will doing so give us better estimates? Does this technique of modeling with Taylor polynomials work on other functions like $\arctan(x)$ and $\ln(x)$? The answer turns out to be yes, at least in some circumstances—including our "bell curve" example. These are some of the deep and important questions explored in a course in real analysis.

11.3.2 Limits and Good Approximations

What makes modeling complex objects with simpler ones work in general? This is a richly layered question with no tidy answer. In each of our two examples, we constructed a sequence of simple objects (polygons, polynomials) designed to approximate a more complex one (a circle, a normal density). In this section we explore what it means for a sequence of functions to converge. Letting $(f_n) = (f_1, f_2, f_3, \ldots)$ be an approximating sequence of functions and f be the function we want to model, the agenda is to have f_n "converge" to f, or become "arbitrarily close" to f as n gets larger. Essentially we want $\lim_{n \to \infty} f_n = f$, but it is not obvious what "close" means in this new context. We know what it means to say two real numbers are close together, but how do we measure the distance between two functions?

A first course in analysis offers a few alternatives. The most basic is called pointwise convergence:

Definition A sequence f_n **converges pointwise** to f on the domain A if the sequence $f_n(x) \to f(x)$ for all $x \in A$. This means that for every $\varepsilon > 0$ and $x \in A$, there exists an $N \in \mathbb{N}$ (perhaps dependent on x) such that $|f_n(x) - f(x)| < \varepsilon$ whenever $n \geq N$. (Abbott's Definition 6.2.1B)

A second, stronger, concept called uniform convergence turns out to be central to deciphering when modeling a complex f with a sequence of simpler f_n functions achieves the desired result:

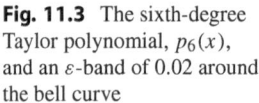

Fig. 11.3 The sixth-degree Taylor polynomial, $p_6(x)$, and an ε-band of 0.02 around the bell curve

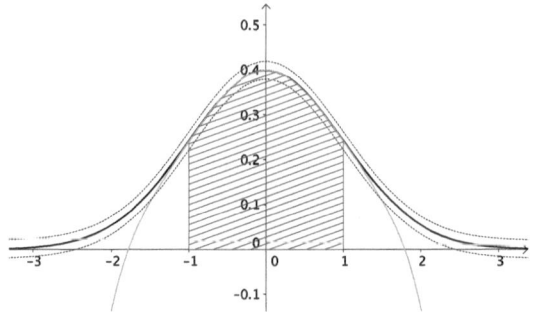

Definition A sequence f_n **converges uniformly** to f on the domain A if for every $\varepsilon > 0$ there exists an $N \in \mathbb{N}$ (independent of x) such that $|f_n(x) - f(x)| < \varepsilon$ whenever $n \geq N$ and $x \in A$. (Abbott's Definition 6.2.3)

A useful way to visualize what is happening in these definitions is to put an ε-band around the limit function f. Figure 11.3 illustrates what this looks like for the "bell curve," $\phi(x) = \frac{1}{\sqrt{2\pi}} e^{-x^2/2}$. Pointwise convergence means that, if we pick an x-value, then the values of $f_n(x)$ eventually enter this ε-band at the particular x-value as n gets larger. Uniform convergence means that there is a point N in the sequence where the entire approximating function f_N, and all the ones to follow, fit completely inside this ε-band. This is what happens with the Taylor polynomials in this example, provided we restrict our attention to the interval $[-1, 1]$. Note how the sixth degree polynomial $p_6(x)$ that we computed earlier is entirely inside the ε-strip on the interval $[-1, 1]$ but slips outside this band for larger values of x. If p_n is the nth degree Taylor polynomial of $\phi(x) = \frac{1}{\sqrt{2\pi}} e^{-x^2/2}$ then $p_n \to \phi$ pointwise on all of $\mathbb{R}$ but the convergence is not uniform on $\mathbb{R}$. Focusing on a smaller domain, however, it is true that $p_n \to \phi$ uniformly on $[-1, 1]$.

Uniform convergence is a vital ingredient if we want to deduce conclusions about the limit function from the approximating sequence. For example, because $p_n \to \phi$ uniformly on this smaller domain, $[-1, 1]$, the sequence of integrals satisfies $\int_{-1}^{1} q_n \to \int_{-1}^{1} \phi$ (Theorem 7.4.4 in Abbott). This means we can compute $\frac{1}{\sqrt{2\pi}} \int_{-1}^{1} e^{-x^2/2}$ to any degree of accuracy we require by integrating the polynomials $p_n(x)$.

Uniform convergence also describes the way the polygons approximate the circle in the earlier example from this chapter. Making use of polar coordinates, we let $P_n(\theta)$ be the point on the regular inscribed n-gon and $C(\theta) = r$ be our original circle. Observing that $P_n(\theta)$ is eventually contained in the shaded ε-strip around $C(\theta)$ in Fig. 11.4 makes it evident that $P_n \to C$ uniformly over the whole domain $[0, 2\pi)$.

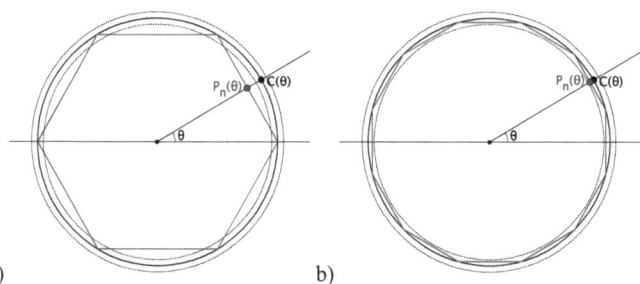

Fig. 11.4 (**a**) A hexagon $P_6(\theta)$, and (**b**) dodecagon $P_{12}(\theta)$, and an ε-band around the circle $C(\theta)$. The largest distance between the circle and the polygon happens at the midpoint of the polygon's side; once n is large enough so that this point is in the ε-band, the rest of P_n is in the band as well

11.3.3 The Real Reason for Analysis

Although it is an important component, $P_n \rightarrow C$ uniformly does not necessarily imply the perimeter of P_n converges to the circumference of C. Drawing conclusions about when the properties of the approximating sequence are passed on to the limit is delicate business in general. In the motivational example at the beginning of his Chap. 6, Abbott invokes phrases such as "closing our eyes to the potential danger of treating an infinite series as though it were a polynomial" (p. 170), "amid all of the unfounded assumptions we are making about infinity" (p. 172), and "despite the audacious leaps in his argument" (p. 173). These statements convey that the modeling of more complex functions with simple ones is complicated.

In our particular Taylor polynomial example things worked out well. If $f_n \rightarrow f$ uniformly on an interval $[a, b]$, then it is always true that the area under f_n converges to the area under f, which means $\int_a^b f_n \rightarrow \int_a^b f$. But it is *not* always the case that the length of f_n converges to the length of f, even when these lengths are properly defined. Figure 11.5a provides a counterexample. Each f_n is composed of segments arranged in a staircase pattern. Shortening the segments increases the number of steps but it does not change the total length of each f_n. The result is a sequence f_n that converges uniformly to a horizontal line segment, but where the lengths of each f_n do not approach the length of the line segment. (In fact the length of each f_n remains a constant, equal to $\sqrt{2}$ times the length of the horizontal segment.)

Fortunately for us, in our inscribed polygons example the lengths of each P_n *do* converge to the circumference of C. (See Fig. 11.5b.) The difference here is that not only does $P_n \rightarrow C$ uniformly but the slopes of P_n approach the slopes of C— which is to say, using our polar notation from before, $P_n'(\theta) \rightarrow C'(\theta)$ (at least at most values of θ). The details of this argument get a bit thorny, and engaging them is not really the point. The main takeaway of this discussion is simply to appreciate the critical role of real analysis as a tool for guiding our intuition about what kinds of conclusions are valid when we approximate complex objects with simpler ones.

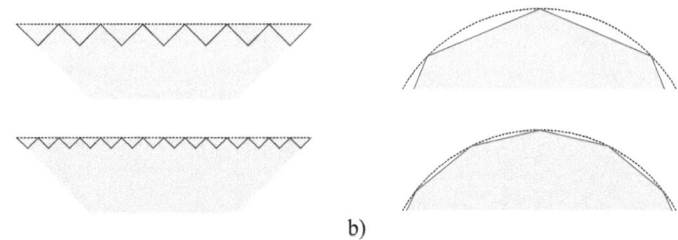

Fig. 11.5 (**a**) "Staircases" approaching a hypotenuse, and (**b**) regular n-gons approaching a circle

11.4   Connecting to Secondary Teaching

The theory of calculus was created through an iterative process of modeling unfamiliar notions with familiar ones. The derivative is defined as a limit of slopes, the integral is defined as a limit of finite sums, infinite series are defined as limits of partial sums, power series are defined as limits of polynomials, and so on. The same principle that is at the center of how the scope of mathematical ideas expand can also be applied to teaching mathematics (TP.4). Teaching is filled with situations where students are confronted with new mathematical ideas or objects. By using simpler, more familiar objects, teachers can build on, and connect to, mathematical ideas students already know as they learn new things.

11.4.1 Modeling the Complex with the Simple as a Pedagogical Practice

Based on the teaching situation posed at the beginning of the chapter, we have been arguing that using regular n-gons to model a circle would be helpful for students in their attempt to decipher the geometry of circles. Consider the following continuation of the teaching scenario:

> Summarizing the students' investigations of circles to this point, Ms. Liu announces:
>
> We measured the circumference of a circle to be between 3 and 4 times the diameter. Let's look at two special cases to understand *why* this would be the case—an inscribed regular hexagon, and a circumscribed square.

(continued)

She then shows students the two figures (see Fig. 11.1), and asks them to compare the length of the perimeters of each polygon to the length of the diameter of the circle. They conclude that the circumference must be greater than $3d$ and less than $4d$. Ms. Liu continues:

The number π is *defined* to be the ratio of circumference to diameter, which means we can use it to find circumference: $C = \pi d$. But what, precisely, is this ratio?

She then directs the students to a dynamic technology applet to explore this ratio with regular n-gons for increasingly large n to better understand π. Doing so further confirms the student's computations, $\pi \approx 3.14$.

Good teaching often leverages concepts and objects students already know to help them investigate ones they don't know. In this scenario, Ms. Liu's follow-up activities use regular n-gons to help her students better understand the relationship between circumference and diameter, but that is not the only benefit. Modeling the complex with the simple is good pedagogical practice in mathematics because it is a reflection of the discipline itself. By using regular n-gons to make inferences about circles, Ms Liu introduces her students to a characteristically mathematical way of approaching problems. Her students are learning about circles in particular and about mathematical modeling in general. They are also getting some hands-on experience with numerical approximation. Perfect precision can be elusive and so approximations, together with information about the quality of the estimates, is a standard and frequently necessary way to solve certain kinds of problems. This issue surfaced specifically in Chap. 3 on irrational numbers, where the idea of modeling the complex with the simple first appeared.

Underlying these many benefits for students is the core idea that teachers should leverage what students already know as resources for learning. When confronted with how to measure the length of a circle's circumference, or why the circumference equals what it does, a good place to begin is with what students already know—like how to find perimeter. Having a comfortable starting point is critical when trying to learn about something unfamiliar.

11.4.2 More with Circles

We've been using inscribed and circumscribed regular polygons to derive conclusions about the circumference of circles. This same strategy can also be used to explore the area. Consider the following three images. The first two use squares; the third uses a dodecagon.

Before moving on, see if you can use these images to come up with lower and upper bounds for the area of the circle in terms of the number of squares with area r^2.

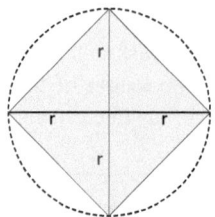 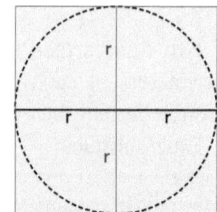 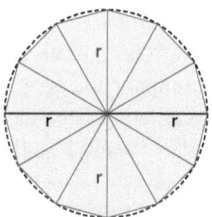

The first two constructions, the inscribed and circumscribed squares, bound the area of the circle between $2r^2$ (from the inscribed square) and $4r^2$ (from the circumscribed square). Visually, each of the $r \times r$ squares is clearly evident in the diagrams. Finding the area of the inscribed dodecagon is more challenging. Figure 11.6 illustrates how four of the isosceles triangles that make up the dodecagon can be rearranged into an $r \times r$ square. (This rearrangement relies on the fact that the central angle for the triangular pieces is $30°$ and that $30°$-$60°$-$90°$ right triangles have familiar side ratios.)

This image helps clarify that the entire dodecagon must have an area of $3r^2$, which underestimates the area of the circle since the dodecagon is inscribed. Letting A be the area of the circle, the circumscribed square and inscribed dodecagon imply

$$3r^2 < A < 4r^2,$$

which points us toward the formula for the area of a circle. Increasing the number of sides in our polygons would increase the precision of our estimates of the area, but is there a way to use this strategy to derive the fundamental relationship $A = \pi r^2$ where $\pi = \frac{C}{d}$?

In a demonstration of his genius, Archimedes achieved this feat by rearranging the isosceles triangles of the approximating regular n-gons into a jagged set of triangular "teeth" which collectively constitute precisely half a rectangle. By choosing n large enough, the base of the rectangle can be made arbitrarily close to the circumference C and the height arbitrarily close to the radius r. The area of the circle is then approximately half the area of this rectangle which is approximately $A = \frac{1}{2}Cr$. Substituting $C = \pi d = 2\pi r$ into this formula yields $A = \pi r^2$.

Investigating complicated shapes via simpler shapes is an effective way to help students justify, and not just memorize, the formula for the area of a circle. Simple dynamic illustrations exist, where some even give students control over how to

Fig. 11.6 A rearrangement of four (of the twelve) isosceles triangles from the inscribed dodecagon

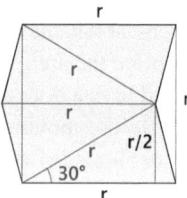

rearrange the pieces to justify the area—turning the investigation into something interactive.

Problems

11.1 Consider the following examples of modeling the complex with the simple, familiar from calculus and real analysis: (i) Using rational numbers to approximate real numbers; (ii) Using secant lines to approximate tangent lines; and (iii) Using rectangles to approximate area under a curve. First, describe how each of these exemplify modeling the complex with the simple. Then, describe why each of these approximation approaches would be considered a "good" way to approximate the desired object. Be detailed in your description of each example, and incorporate mathematical language from analysis.

11.2 Consider the following statement: "A line tangent to a circle is exactly perpendicular to the radius at the point of tangency." This is a true statement, and one taught in high school geometry. As a teacher, how could you use the fact that one can use secant lines to approximate tangent lines in order to explain why tangents on a circle (see below) are exactly perpendicular to the radius (at the point of tangency)? You might consider several example secant lines.

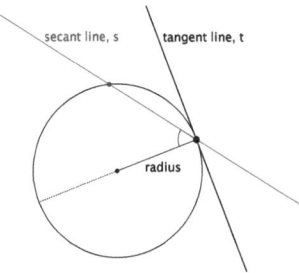

(Note: Either of the following two theorems might be of help: (i) The measure of an inscribed angle is half the angle of its intercepted arc; or (ii) Two radii form an isosceles triangle with the center.)

11.3 Archimedes also used a modeling the complex with the simple approach in his determination of a sphere's volume. He did so by modeling a sphere with cones and frustrums—a frustrum is the bottom "slice" of a cone. Their volumes are:

- $V_{cone} = \frac{1}{3}\pi h r^2$, where r is the radius of the cone's base and h is the cone's height;
- $V_{frust} = \frac{1}{3}\pi h (R^2 + r^2 + R \cdot r)$, where R is the radius of the frustrum's base, r is the radius of the frustrum's top, and h is the frustrum's height.

Compute the volume of the first three spherical approximations, which come from the three 2D images depicted below. (You will need to imagine rotating the shaded 2D regions around the horizontal x-axis to form the relevant cones and frustrums.) What does each approximation give as a lower-bound for the volume of a sphere, which we know to be $V_{sph} = \frac{4}{3}\pi r^3$? (Hint: you will need to solve for the radius of the top of each frustrum using the additional dotted segments.)

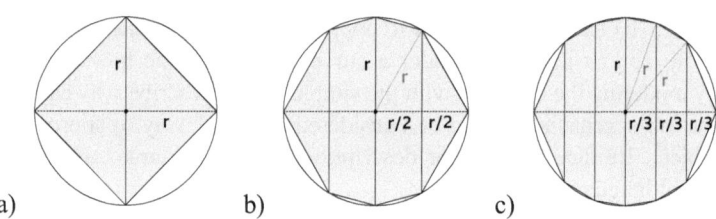

a)   b)   c)

11.4 Statisticians commonly investigate relationships between bivariate data. That is, whether the data suggest a relationship between two quantitative variables (often plotted on an x- and y-axis). (i) A common approach is to try to fit a linear model to the data. Describe why you think statisticians often use a *linear* model; make sure to discuss it in relation to TP.4. (ii) To determine how good of a fit a linear model is, a common approach is to compute deviations—i.e., the distance between an observed data point's y-value, and the one a linear model would predict based on its x-value. This measures the distance from a (data) point to a (best-fit) line—and measures distance differently than how we would in geometry (see Problem 6.3). In terms of modeling the relationship between bivariate data, why might statisticians choose to use this measure of distance?

11.5 Describe why, based on this chapter, it might be important for secondary students to understand polynomials. Explain how you might motivate a class of secondary students to learn about polynomials.

11.6 Abbott [1] (p. 171) describes the identity, $\frac{1}{1+x} = 1 - x + x^2 - x^3 + x^4 \ldots$, as being a special case of Newton's generalization of the Binomial Theorem. Prior to giving the identity, he specifies that to Newton the infinite series was "meaningful, at least for $x \in (-1, 1)$." One could consider this statement as related to TP.1. Describe how Newton's observation (as summarized by Abbott) is related to TP.1. Explain what happens to the identity outside the scope of $x \in (-1, 1)$, as well as why that happens.

11.7 This chapter looked more closely at Taylor polynomials. When describing Taylor's formula for determining the coefficients, Abbott [1] states: "the idea is to assume that f has a power series expansion and deduce what the coefficients must be" (p. 199) and, later on that page, "To derive Taylor's formula, *we assumed that* *f actually had a power series representation*" (italics added). Then, Abbott asks, "But what about the converse question?" (p. 200). First, discuss Abbott's comments and questions in relation to TP.3. Specifically, give a logically precise statement of Taylor's formula, and give a logically precise statement of the converse. Then, discuss the converse statement in light of the counterexample Abbott gives (on p.

$$203): g(x) = \begin{cases} e^{-1/x^2} & \text{for } x \neq 0 \\ 0 & \text{for } x = 0 \end{cases}.$$

11.8 Consider Examples 6.2.2(ii) and 6.2.2(iii) from Abbott's text. They are: (ii) $g_n(x) = x^n$ on the set $[0, 1]$; and (iii) $h_n(x) = x^{1+\frac{1}{2n-1}}$ on the set $[-1, 1]$. You might look at the depictions in the text [1, p. 175], or graph the first few of the functions in those sequences to get a feel for each example. Consider the examples in light of TP.2. Identify which is related to exploring whether $\lim(f_n') = (\lim f_n)'$, and which is related to examining properties of f_n and properties of $\lim f_n (= f)$. Explain your reasoning.

References

1. Abbott, S. (2015). *Understanding analysis* (2nd ed.). New York, NY: Springer.
2. Poyla, G. (1945) *How to solve it.* Princeton, NJ: Princeton University Press.

The Riemann Integral and Area-Preserving Transformations

12

12.1 Statement of the Teaching Problem

Mathematics requires rigorous justification for its propositions. A challenge in teaching mathematics is presenting students with proofs that also provide conceptual insights (TP.5). Doing so requires that the instructor have the dexterity to think about a problem in different ways and determine when one approach might be more effective than another.

One area where this flexibility is especially valuable is generating the formulas for area of figures like triangles, parallelograms, circles, and ellipses. The notion of area starts as an enumeration of *square units* and a consequence of this bedrock idea is that a rectangle has area $A_{rect} = lw$. A standard strategy for determining the area of shapes such as triangles and parallelograms is to cut these new shapes into pieces that can be reassembled into a rectangle—or into some other planar shape whose area has already been determined.

Consider the following pedagogical situation:

> A geometry teacher, Mr. Williams, shows the class how to cut the triangle off the left side of a parallelogram and move it to the right to see that the area of a parallelogram is given by $A_{par} = bh$.

(continued)

© The Author(s), under exclusive license to Springer Nature Switzerland AG 2022
N. H. Wasserman et al., *Understanding Analysis and its Connections to Secondary Mathematics Teaching*, Springer Texts in Education,
https://doi.org/10.1007/978-3-030-89198-5_12

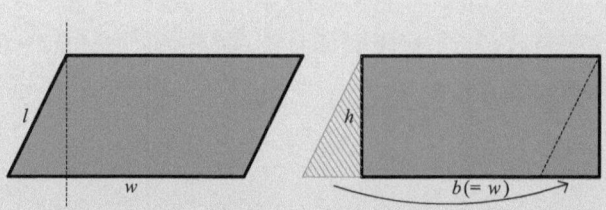

Mr. Williams then decides to have the class look at the "oblique" parallelogram below, created by sliding the top segment of the original parallelogram
quite far to the right while maintaining the base w and the height h.

Mr. Williams starts to use the "cut-reassemble" argument he used before
to find the new area but gets stuck because moving the sliced off triangle no
longer makes a rectangle. He then then wonders if there is another way to
explain to his students that the area of this parallelogram is still $A_{par} = bh$.

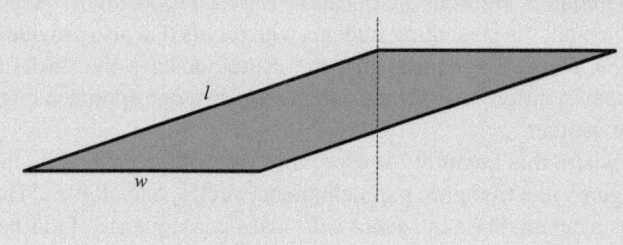

In the first example, the "cut-reassemble" algorithm results in a rectangle with
dimensions $w \times h_w$ where h_w is the height of the parallelogram when we take w as
the base. This yields the area formula $A_{par} = wh_w = bh$. The problem comes in the
second example where the line cutting the original parallelogram no longer meets
the base. The same area formula certainly holds, but we are momentarily without
a proof of this fact. (Some might be surprised to read that the areas of the three
figures above are all the same!) This is a situation where the teacher would benefit
from having multiple explanations for the concept at hand (TP.6). Figuring out how
the argument might be modified or what other justifications might exist requires us
to think flexibly about area-preserving transformations.

Before moving on, think about how you might adapt the "cut-reassemble"
argument to this "oblique" parallelogram, or if there is some other approach you
could use to justify the area formula.

12.2   Connecting to Secondary Mathematics

12.2.1 Problematizing Teaching and the Pedagogical Situation

One solution for dealing with the "oblique" parallelogram is to change the direction of the line cutting it. The base of the figure need not be orthogonal to the page (as shown below).

It turns out it will always be the case that one or the other cut in a (non-rectangular) parallelogram will intersect the base.[1] Although this is a way to salvage the "cut-reassemble" approach, students may have trouble recognizing that a different cut could work and that the base need not be orthogonal to the page. Another drawback is that the resulting rectangle has different dimensions than the one we got from before. The non-vertical cut produces a rectangle with dimensions $l \times h_l$ rather than $w \times h_w$, obscuring the interesting fact that both parallelograms have the same area.

A second option is to use several "cut-reassemble" steps.

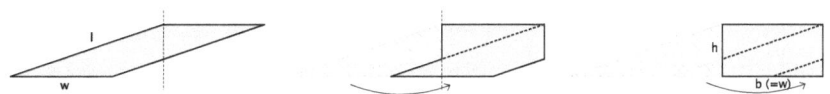

Although this produces a valid argument, the multiple steps add to the complexity and mask any conceptual insight about *why* the areas of the two parallelograms come out equal. The proof works but it feels a bit like forcing a square peg into a round hole. The "cut-reassemble" approach depicted in the teaching situation gives the impression that only *one* cut is necessary. By giving students justifications that are applicable only to specific diagrams, we may be encouraging them to see proof methods as not applicable to the entire classes of objects the diagrams are intended to represent.

---

[1] For w and l of the (non-rectangular) parallelogram, and the *acute* angle θ between them ($0 < \theta < \pi/2$), the vertical cut (with w as the base) will meet the base so long as $l \leq w \sec \theta$. If this is not the case then $l > w \sec \theta \implies l > w \cos \theta$, since $\sec \theta > \cos \theta$ on $0 < \theta < \pi/2$, which means $w \leq l \sec \theta$, and so the non-vertical cut (with l as the base) *must* meet the base.

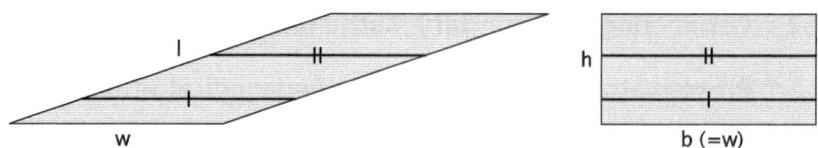

Fig. 12.1 Cavalieri's (2D) principle justifying the area of the "oblique" parallelogram in relation to a $w \times h_w$ rectangle

12.2.2 Area-Preserving Transformations and Cavalieri's Principle

The "cut-reassemble" approach to finding area formulas is based on the idea of *area-preserving transformations*. Because we move the pieces of the original region in a rigid way that does not alter their shape, putting the pieces back together (without gaps or overlaps) creates a different region with the same area. It is this area-preserving nature that allows us to determine the area of an unknown region in terms of the area of a known one. (This point should recall TP.4—modeling more complex objects with simpler ones.) But there are other area-preserving transformations that provide other means to justifying areas.

The Common Core State Standards in Mathematics—which currently guide the content taught to U.S. secondary school students—contains the following standard: *"Give an informal argument using Cavalieri's principle for the formulas for the volume of a sphere and other solid figures"* [2]. In this instance, Cavalieri's principle is being invoked to solve a three dimensional problem about volume, but it is equally valid in two dimensions where it can be used in conjunction with area:

Cavalieri's (2D) Principle Suppose two regions in a plane lie between two parallel lines in that plane. If every line parallel to these two boundary lines intersects both regions in line segments of equal length, then the two regions have equal areas.

Cavalieri's principle asks us to conceptualize a region in the plane as being "composed" of a stack of parallel line segments. If we have another region "composed" of line segments with the same lengths, this new region should have the same area as the original. As a first example, Cavalieri's principle offers a quick and efficient way to conclude that the areas of the parallelogram and rectangle (in Fig. 12.1) are equal, based on the fact that the line segments made by the parallel lines are always congruent. Notice how this might relate to the dilemma in the teaching situation.

Although the statement of Cavalieri's principle has a static feel—comparing lengths of line segments—a more dynamic interpretation can be given in terms of a specific kind of transformation that we make precise:

Definition Consider a region in the plane composed of parallel line segments. A **segment-skewing transformation** is one that translates each line segment along the line to which it belongs in such a way that the composite transformation yields a new region in the plane.[2]

With this new vocabulary, we can restate Cavalieri's principle by saying that *a segment-skewing transformation is area-preserving.* Analogous to the "cut-reassemble" technique where we move the pieces of the original in an area-preserving way, we can imagine transforming the "oblique" parallelogram in Fig. 12.1 into the rectangle by applying a segment-skewing transformation that slides each horizontal segment the proper distance to the right. Cavalieri's principle tells us the area of the transformed region is the same as the original.

12.3 Connecting to Real Analysis

The primary connection between Cavalieri's principle and real analysis is though the theory of integration. Specifically, we look at how the Riemann integral can be used as justification for this principle, as well as how Cavalieri's principle provides some geometric insight into various integration rules.

12.3.1 Justification for Cavalieri's Principle via Integration

Why is Cavalieri's principle true? Considering a planar region to be composed of line segments introduces legitimate conceptual difficulties. Unlike the pieces in the "cut-reassemble" approach, these individual line segments have no area on their own yet come together to create a region that does possess an area. At issue is how to define what we mean by area for planar regions that aren't rectilinear, and this is where the Riemann integral can offer us some assistance.

Let f be a continuous function taking positive values on the interval $[a, b]$. Intuitively, the integral $\int_a^b f(x)dx$ represents the area of the region S bounded by $x = a$, $x = b$, $f(x)$ and the x-axis (see below). To compute this rigorously we can set $\Delta x = \frac{b-a}{n}$, $x_i = a + i\Delta x$ for $i = 1, 2, \ldots, n$ and conclude:

$$\text{Area}(S) = \int_a^b f(x)dx = \lim_{n \to \infty} \sum_{i=1}^n f(x_i) \cdot \Delta x.^3$$

---

[2] We note that a segment-skewing transformation is related to, but more flexible than, a *shear* transformation.

[3] This calculation of Area(S) is not the definition of the Riemann integral but a consequence of it when the function is continuous. The treatment of the integral in Abbott employs upper and lower sums. The quantity in this expression is a Riemann sum evaluated for a specific tagged-partition. Knowing that the integral exists, it follows that this particular sequence of approximations must converge to the proper value. See Abbott, Theorem 8.1.2.

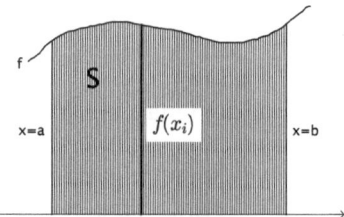

As with all of the other formulas for area in this chapter, the fundamental ingredient for defining the integral is the area of a rectangle; in this case $f(x_i) \cdot \Delta x$. The Riemann integral defines 'Area(S)' to be the limit of the sums of increasingly more, and increasingly thinner, rectangles. To connect this to Cavalieri's principle, imagine these tall thin rectangles as looking more and more like the parallel line segments from Cavalieri's principle, but this time oriented vertically.

Now let's apply a segment-skewing transformation to S (in the vertical direction) by adding a continuous function $t(x)$ to $f(x)$. That is, we'll let $g(x) = f(x) + t(x)$ and define the new region T between $x = a$ and $x = b$ to be the one bounded above by $g(x)$ and below by $t(x)$. Because $f(x) = g(x) - t(x)$, the corresponding vertical segments that make up S and T have the same length. Cavalieri's principle then asserts that Area(S) = Area(T), and in this setting we have the Riemann integral to help us look under the hood to see why this is the case. Specifically,

$$\text{Area}(S) = \lim_{n\to\infty} \sum_{i=1}^{n} f(x_i) \cdot \Delta x = \lim_{n\to\infty} \sum_{i=1}^{n} [g(x_i) - t(x_i)] \cdot \Delta x = \text{Area}(T).$$

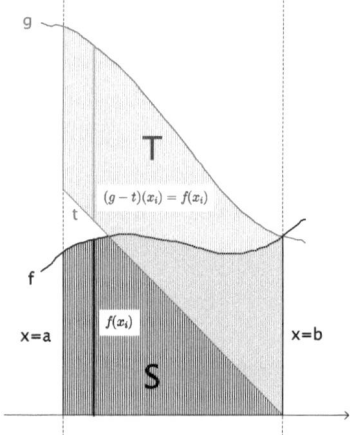

With a little polish, this example can be upgraded to a full proof of Cavalieri's principle in two dimensions. All we need to do is let S be more general by allowing it to be bounded from below by an arbitrary continuous function (rather than the x-axis). Restricting our attention to continuous functions seems reasonable since all these curves are meant to be the boundaries of planar figures like parallelograms and ellipses. The assumption of continuity ensures the functions are Rieamnn integrable (see Theorem 7.2.9 in Abbott [1]) and the rest is smooth sailing.

12.3.2   Cavalieri's Principle and Integral Properties

Having used the theory of the Riemann integral to ground Cavalieri's principle on a solid mathematical footing, we can flip the script and see how Cavalieri's principle provides geometric insight for some important properties of the integral.

Consider the property

$$\int_a^b (f - g) = \int_a^b f - \int_a^b g.$$

This equation expresses a relationship between *three* functions, $f, g, (f - g)$, and the areas under their graphs (below). (We presume $f(x) > g(x)$ for all $x \in [a, b]$.)

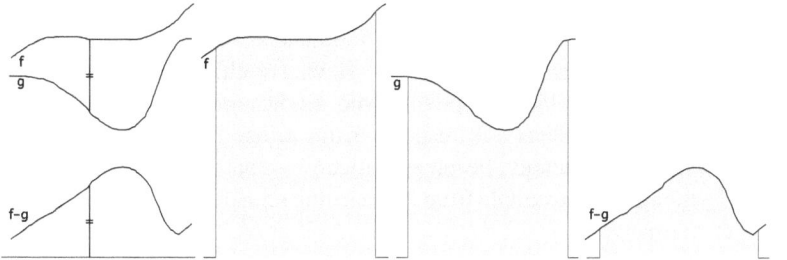

While this rule conveys a fact about the algebra of integrals, the geometric justification amounts to a segment-skewing transformation. Specifically, translate each vertical segment in the original planar region between f and g down to the x-axis (see Fig. 12.2). The boundary curves of this new region are the function $f - g$ and the x-axis. The area-preserving nature of the transformation means the area of the first region, $\int_a^b f - \int_a^b g$, must be equal to that of the second, $\int_a^b (f - g)$.

Cavalieri's principle helps us understand why this integral property is sensible, from a geometric and not just algebraic standpoint. Other integral properties such as

$$\int_a^b kf = k \int_a^b f \quad \text{or} \quad \int_a^b f + \int_a^b g = \int_a^b (f + g)$$

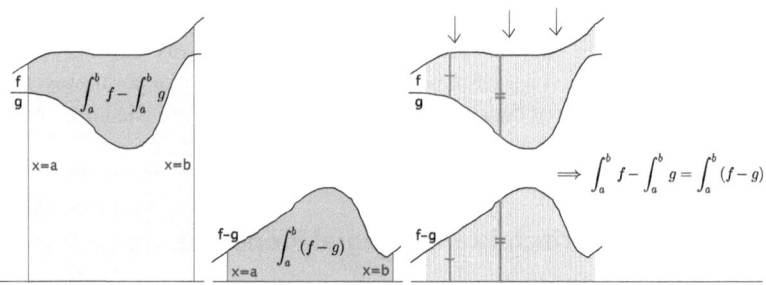

Fig. 12.2 Segment-skewing the area between functions f and g to form the area under $f - g$

have similar geometric justifications in relation to segment-skewing transformations. They each rely on recognizing a region in the plane as having an area equal to some other region(s). (Problems 12.6 and 12.7 ask you to look at these.)

12.4   Connecting to Secondary Teaching

Although the "cut-reassemble" algorithm can be used to justify the area formulas for many shapes treated in secondary school mathematics, the segment-skewing approach has its particular benefits. One rationale is simply knowing a variety of approaches to a given problem (TP.6). Having multiple explanations affords teachers the ability to relate to students who might require a different approach to a problem. Some students will respond better to one method than another! On occasions where a justification becomes difficult in a particular situation—like the "oblique" parallelogram example from the teaching scenario—it is always important to have an alternative.

Incorporating the segment-skewing approach to area also provides points of connection to other content. Cavalieri's principle is typically used for finding the volume of three-dimensional solids. Becoming familiar with Cavalieri's principle in the plane provides a foundation for appreciating its use in use in three-dimensional space. Even better, segment-skewing foreshadows fundamental ideas in calculus; namely, how the Riemann integral approaches area. This way of thinking about area is not just different, it is productive. Exposing students to Cavalieri's principle lays a groundwork on which they can eventually build an understanding of calculus.

12.4.1 Areas of Polygons

Let's return to the "oblique" parallelogram in the initial teaching situation and suppose the teacher proceeds as follows:

Mr. Williams acknowledges that with the "oblique" parallelogram, the "cut-reassemble" approach would take multiple steps. So instead he asks students to think about its area in the following way:

First, I want you to imagine a bunch of pencils lined up that fill in the "oblique" parallelogram's area (like in Fig. 12.1).

If we push those pencils sideways to create a new shape (like in Fig. 12.1), then we haven't changed the area. Doing so produces a rectangle that is $w \times h_w$, and so the area is still $A_{par} = A_{rect} = w h_w = bh$.

Mr. Williams's use of a segment-skewing transformation provides a relatively simple explanation for why the area of the "oblique" parallelogram is still $A_{par} = bh$. Moreover, it provides another means by which to transform the parallelogram into the same $w \times h_w$ rectangle as had been accomplished with the prior "cut-reassemble" approach. This resolution aligns with TP.5 because it provides a justification for the area formula in this unusual case, and it aligns with TP.6 because it provides an alternative explanation. Introducing students to segment-skewing transformations gives them an alternate method for conceptualizing and justifying the areas of planar figures.

To sharpen Mr. Williams's pencil analogy, consider the parallelogram bounded between $x = 0$, $x = h$, $f(x) = mx + w$, and $g(x) = mx$. Subtracting $g(x)$ transforms the parallelogram into a rectangle with height $w = f(x) - g(x)$, and Cavalieri (or, equivalently, properties of the integral) does the rest.

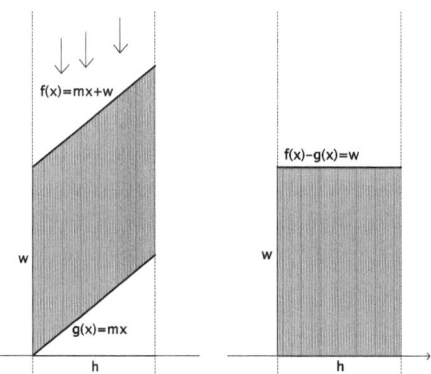

Cavalieri's principle can also be used to study other polygons. A segment-skewing transformation can transform any triangle into a right triangle with the same base and height, thereby confirming the area formula $A_{tri} = \frac{1}{2}bh$. This exercise also provides students with another way to understand what is meant by the "height" or "altitude" of a triangle.

To determine the area of a trapezoid we can use a combination of segment-skewing and the "cut-reassemble" algorithm. As a first step we apply a segment-skewing transformation in the direction parallel to the base so that we end up with a right trapezoid. Then we split the right trapezoid into a rectangle and a right triangle to compute

$$A_{trap} = A_{rec} + A_{tri} = b_1 h + \frac{1}{2}(b_2 - b_1)h = \frac{1}{2}b_1 h + \frac{1}{2}b_2 h.$$

The punchline is that Cavalieri's principle is a valuable tool in the toolkit, for teachers and students. It provides additional opportunities for reasoning about geometric shapes and offers another way of thinking about area and area-preservation that is important for more advanced mathematical studies.

12.4.2 Areas of Ellipses

As one final class of examples, consider the case of an ellipse. The area of an ellipse is given by $A_{ellip} = \pi ab$ where a and b are the two axes. This formula is less familiar to students, and most textbooks give little insight as to why it is sensible. Cavalieri's principle provides the sort of conceptual insight that is frequently missing.[4]

We start with a circle of radius 1, which we know has an area of π square units. Suppose our ellipse has a major axis of 4 in the x direction and a minor axis of 2 in the y direction. Starting with the unit circle, apply a horizontal dilation with a factor of 4 to form the major axis. Stretching horizontally creates three new sets of crescent-shaped areas, each of which by Cavalieri's principle has an area equal to the original unit circle (Fig. 12.3a). (To visualize this, imagine the original unit circle cut in half vertically and composed of horizontal line segments; push those outward to create the two adjacent crescent-shaped regions.) Now that we have stretched along the major axis, we can stretch the resultant region in the vertical direction by a factor of 2 to form the minor axis. Vertically stretching this region, which has an area of 4π, creates a set of crescent-shaped areas that has an area equal to the original region (Fig. 12.3b). Hence, $A_{ellip} = (4\pi) \cdot 2 = 8\pi$. This argument can be generalized to an arbitrary a and b and provides the needed insight into why the area of an ellipse is $A_{ellip} = \pi ab$.

---

[4] We restrict our discussion to natural numbers a and b, although the argument can be extended to the real numbers with a slight modification to Cavalieri's principle.

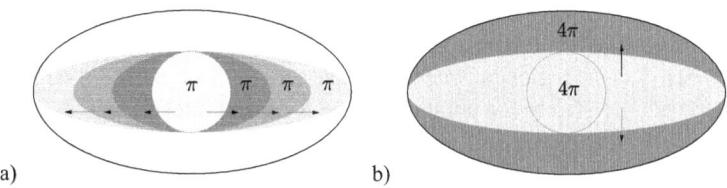

Fig. 12.3 The area formula for an ellipse via segment-skewing (**a**) horizontally, and (**b**) vertically

12.4.3 Transformations That Are Not Continuous

One condition of Cavalieri's principle was that the composite transformation yield a new planar region. But will this always be the case? What happens, for example, when the segment-skewing transformation applied to a region is not continuous? Does it make sense to discuss Cavalieri's principle with respect to sets in the plane created in this way? Although we get a well-defined collection of points when we translate by a discontinuous transformation, it's not so obvious whether this new set has a well-defined area. Let's probe this condition further (TP.1).

Consider the planar region S depicted previously that lies under the function $f(x)$. Now translate S by adding the function

$$t(x) = \begin{cases} 0 & \text{if } x \in [a, c) \cup (c, b] \\ 1 & \text{if } x = c \end{cases}.$$

Because $t(x)$ is zero everywhere except at $x = c$, our new set of points looks like S with the exception of a single segment raised up by one unit.

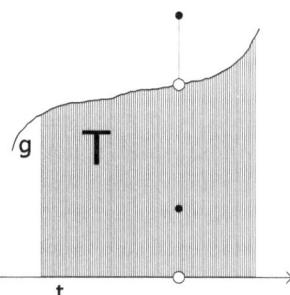

The geometric object does not constitute a planar region, which means Cavalieri's principle does not technically apply. Yet, it is reasonable to think that if we *were* to assign an area to this new object, it should be the same as the area of the original region. It turns out, Riemann integration does just this. Unlike the derivative, which is not defined at a point of discontinuity (see Chap. 9), the Riemann integral is defined for functions with a discontinuity.

Recall that S is the region between $f(x)$ and the x-axis. Letting T be the new set obtained by adding $t(x)$ to the points of S, the Riemann integral assigns T the area $\int_a^b g(x)dx - \int_a^b t(x)dx$ where $g(x) = f(x) + t(x)$. By properties of the integral,

$$\text{Area}(T) = \int_a^b g(x)dx - \int_a^b t(x)dx = \int_a^b (f+t)(x)dx - \int_a^b t(x)dx =$$

$$= \int_a^b f(x)dx + \int_a^b t(x)dx - \int_a^b t(x)dx = \int_a^b f(x)dx = \text{Area}(S).$$

That is to say, the Riemann integral assigns the new geometric object T the same area as S, thereby *expanding* the class of segment-skewing transformations that can be considered area-preserving. This works because $t(x)$ is a Riemann integrable function on $[a, b]$ even though it has a point of discontinuity. When a non-Riemann-integrable transformation is applied to S—such as Dirichlet's function restricted to $[a, b]$—we cannot use the Riemann integral to find the resulting area. This leads to a curious state of affairs: although we might instinctively feel this new unruly set of points in the plane should still have area S (since it is composed of translated vertical line segments), Cavalieri's principle doesn't apply and the Riemann integral can no longer help. Not all segment-skewing transformations produce geometric objects for which the Riemann integral assigns the same area. Shortcomings of this nature in the Riemann integral eventually led to new definitions of integration, each motivated in part by the desire to create a mathematically sound definition of area applicable to larger classes of sets. (For further discussion of the strengths and weaknesses of the Riemann integral as well as an introduction to some alternative integrals, see Abbott's Section 7.6, 7.7, and 8.1.)

Problems

12.1 In this chapter, we explored an "oblique" parallelogram. Consider an "oblique" triangle—a triangle whose vertex is not over the base (i.e., the altitude falls outside the triangle). First, try to produce a "cut-reassemble" argument for why the area of an "oblique" triangle is still $A_{tri} = \frac{1}{2}bh$. Second, give a segment-skewing argument for its area. In both cases, make sure to connect each argument all the way back to the area of a rectangle.

12.2 One way that a teacher could justify the area formula for a kite, $A_{kite} = \frac{1}{2}d_1d_2$ (with d_1 and d_2 being the lengths of the two diagonals) is to use a "cut-reassemble" transformation to turn the kite into a rectangle with base d_1 and height $\frac{1}{2}d_2$. As an alternative, a segment-skewing transformation could be used to turn the kite into a triangle, with base d_1 and height d_2. Provide an explanation for the area formulas for a kite that you might give as a geometry teacher using this approach; make sure to include a description for how this transformation gives a justification of the area formula for a kite.

12.3 Stretch a semicircle vertically by a factor of two. How does the crescent-shaped area that was added by this stretching compare to the original semicircle's area? Give a detailed explanation for your reasoning.

12.4 On an international test, a problem similar to this one appeared:

> Farmer Joe and Farmer Bob share a fence between their properties (see figure). Currently, their properties are exactly equal in terms of area. However, they would like to create a new fence—one that is straight—but that retains their properties having equal areas. How might you determine a straight fence between their properties that would not result in the loss of any area for either farmer?

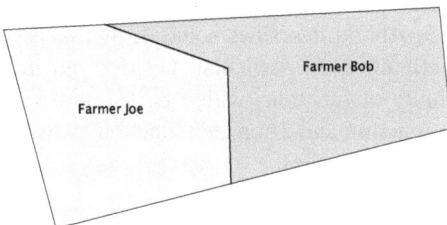

Use a "segment-shifting" transformation to determine how to construct the straight fence in a way that preserves the area of the two properties.

12.5 Transform a circle of radius 1, first, by segment-skewing in the vertical direction (until you have a flat base) and then, second, by segment-skewing the result of your first transformation in the horizontal direction. What is the resultant shape? Use this transformation to provide a justification that could be used by a geometry teacher for the fact that a circle with double the radius (i.e., 2:1) has an area that is four times the original's area (i.e., 4:1). (More generally, with similar planar figures, if the ratio of side lengths is $a : b$, the ratio of areas is $a^2 : b^2$.)

12.6 Provide a geometric justification that you might give as a calculus teacher, connected to segment-skewing transformations and Cavalieri's principle, for the following integral rule: $\int_a^b kf = k \int_a^b f$ (consider only the case for $k \in \mathbb{N}$). [Presume $f : [a, b] \rightarrow \mathbb{R}^+$]

12.7 Provide a geometric justification that you might give as a calculus teacher, connected to segment-skewing transformations and Cavalieri's principle, for the integral rule: $\int_a^b f + \int_a^b g = \int_a^b (f + g)$. [Presume $f, g : [a, b] \rightarrow \mathbb{R}^+$]

12.8 Consider the triangular region S between the lines $x = 0$, $x = h$, $f(x) = (m - \frac{b}{h})x + b$, and $g(x) = mx$ (for $m, b, h \in \mathbb{R}^+$). (i) Provide an algebraic approach (relying on integral properties from calculus) to find the area of S; (ii) Provide a geometric justification for the area of S based on a segment-skewing transformation.

12.9 Draw an arbitrary quadrilateral $ABCD$. Explain at least two ways that you could transform the quadrilateral $ABCD$ into a triangle, while preserving the area. That is, how to create a triangle that has an area equal to the drawn quadrilateral.

12.10 Consider a regular hexagon and a regular octagon, with side length s. Provide a segment-skewing transformation that would turn them both into trapezoidal-like figures (one is a trapezoid, the other a trapezoid on top of a rectangle). Find the appropriate dimensions (you'll need to use some trigonometry), and then determine the areas of the regular hexagon and regular octagon using the fact that $A_{trap} = \frac{1}{2}h(b_1 + b_2)$.

12.11 In his Sect. 7.3, Abbott describes some of the nuances around integrating functions with discontinuities. In particular, he gives an inductive argument for the Riemann-integrability of functions with a *finite* number of discontinuities. He then gives Dirichlet's function and Thomae's function as two cases with an *infinite* number of discontinuities. He writes:

> ...so we conclude that Dirichlet's function is *not* integrable...[but] we should realize that Dirichlet's function is discontinuous at *every* point in [0, 1]. It would be useful to investigate a function where the discontinuities are infinite in number but do not necessarily make up all of [0, 1]. Thomae's function...is one such example...[and] we will see that Thomae's function *is* Riemann-integrable...

First, describe how Abbott's use of these two examples is related to TP.2. What mathematical purpose did each serve? Second, in the last section of this chapter we looked at one segment-skewing transformation t that had a *singular* discontinuity. Describe what the conclusions from Abbott's examples mean for the kinds of segment-skewing transformations t that would still be area-preserving using Riemann integration.

References

1. Abbott, S. (2015). *Understanding analysis* (2nd ed.). New York, NY: Springer.
2. Common Core State Standards in Mathematics (CCSSM). (2010). Retreived from: http://www. corestandards.org/the-standards/mathematics.

The Fundamental Theorem of Calculus and Conceptual Explanation

13.1 Statement of the Teaching Problem

The centerpiece of the theory of calculus is the discovery that finding the instantaneous rate of change (the derivative) and computing the area under a curve (the integral) are inverse processes. The Greeks developed a notion of infinitesimals and could use them to calculate area. Tangent lines and instantaneous rates of change became a topic of study in the first half of the seventeenth century. A full recognition that the tangent line problem and the area problem were intertwined in a powerful way happened gradually over the ensuing decades and is now appropriately referred to as the Fundamental Theorem of Calculus (or FTC).

The Fundamental Theorem provides a straightforward algebraic algorithm for solving the complicated geometric problem of computing the area under the curve. This indispensable tool is at the heart of the reason why calculus transformed the evolution of science. But as students master the steps for how to use the FTC to compute definite integrals, we also want them to have sense of why the theorem works. The long historical path to its discovery suggests this is a tall order. It's not at all obvious that finding an anti-derivative should be helpful in computing an area.

Consider the following pedagogical situation:

> A calculus teacher, Ms. Kalili, is giving an introduction to the FTC and explaining what the theorem means. She says, "essentially, integrals and derivatives are inverses," and then illustrates with an example while interacting with the class:

(continued)

© The Author(s), under exclusive license to Springer Nature Switzerland AG 2022
N. H. Wasserman et al., *Understanding Analysis and its Connections to Secondary Mathematics Teaching*, Springer Texts in Education,
https://doi.org/10.1007/978-3-030-89198-5_13

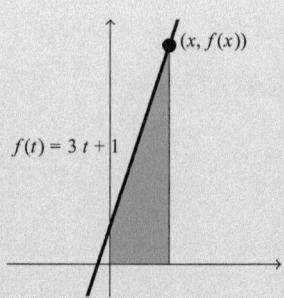

Ms. Kalili: For example, if we take this function, $f(t) = 3t + 1$, and we want to take the integral from 0 to some number x, what would the integral be?

Destiny: The integral would mean the area under the line, which forms a trapezoid. Let's see. The two bases of the trapezoid would be 1 and $f(x) = 3x + 1$, and the height would be x. So the area should be $\frac{1}{2}x \cdot (3x + 1 + 1) = \frac{3}{2}x^2 + x$.

Isabela: Interesting, I saw the area as a rectangle, with a triangle on top. The rectangle has an area of $1 \cdot x$, and the triangle's area would be $\frac{1}{2}x \cdot 3x = \frac{3}{2}x^2$. So the total area would be $\frac{3}{2}x^2 + x$, which is the same!

Ms. Kalili: Very good! You all just found the integral of f from 0 up to a point x. Now, let's take the result, $\frac{3}{2}x^2 + x$, and take its derivative. What do you get?

Destiny: Wait a minute, this is interesting. It would be $2 \cdot \frac{3}{2}x + 1 = 3x + 1$, which is essentially the function we started with!

Ms. Kalili: That's correct! Notice that by taking the integral, and then the derivative, we have gone back to our original function f. This will always happen.

Ms. Kalili used a particular example function to explore the FTC relationship, doing so both geometrically and algebraically. Her students had a working definition of the integral as the area under the curve. By using the linear function $f(t) = 3t + 1$, Ms. Kalili created an example where her students could compute the integral precisely, and in multiple ways, using familiar techniques of geometry. Having an example with multiple entry points allowed them to focus on the more abstract issue, which is that computing areas is somehow related to derivatives and anti-derivatives—a positive instance of TP.2. Yet, while the example serves a reasonable purpose, there are also other considerations to be made in teaching. Ms. Kalili's goal with this example is to unearth, explain, and justify the relationship FTC describes.

How well does this example advance that agenda? Before moving on, think about what you, as a teacher, might do next.

13.2 Connecting to Secondary Mathematics

13.2.1 Problematizing Teaching and the Pedagogical Situation

A recurring theme of this book is the distinction between following algorithms and reasoning conceptually. Closely related to this is the distinction for teachers between providing justifications that substantiate and those that offer useful intuition for why a theorem is true. Ms. Kalili's example provides a good first step to appreciating that the integral is an anti-derivative, but there is nothing yet about why this is the case. The teacher's example highlights the FTC as a rule without explaining why the rule works. Moreover, the explanation the teacher gives is based on just one example, which is not sound justification in the general case (e.g., "This will always happen"). Good teaching certainly involves helping students learn procedures, but it also requires giving them them an understanding of where the rules come from. This distinction between procedural and conceptual is at the heart of what we discuss in this chapter. The Fundamental Theorem of Calculus is a flashpoint for this challenge because students often focus on the mechanical steps of computing integrals so intently that finding an anti-derivative becomes their de facto definition of what a definite integral is.

We offer an examination of FTC meant to address this challenge.

13.2.2 Rate of Change

As we discussed in Chap. 9, the applied or physical interpretation of the derivative is the instantaneous rate of change. The formal definition,

$$F'(x) = \lim_{x \to c} \frac{F(x) - F(c)}{x - c}$$

combines a limit with the familiar formula for slope. If we think in terms of measurement—if F represents some physical quantity like population or volume, whose value depends on some other quantity x such as time or distance—then the slope formula measures a change in F based on an incremental change in x. Each 1-unit change in the x quantity corresponds to a particular change in the F quantity. The units are essential to interpreting the meaning. If F is the population of the earth and x is the year, the rate of change units are people per year. If F is the cost in dollars to manufacture x bicycles, then the rate of change units are dollars per bike.

The expression $\frac{F(x) - F(c)}{x - c}$ gives the average rate of change of F with respect to x over the interval from x to c which usually makes sense in whatever applied situation we consider. Taking the limit as $x \to c$ gives the instantaneous rate of change, which has the same units but requires a little more care to properly interpret because it may not describe something meaningful in terms of an attribute's measurement.

13.2.3 Area Conceptualized as in Cavalieri's (2D) Principle

Cavalieri's principle, discussed in Chap. 12, was a precursor to calculus that helped lay the intuitive groundwork for the integral. It was an early tool for computing areas of shapes with curved boundaries. As an example, consider the region S between a continuous function $f(x)$ and the x-axis over the interval $[a, b]$ pictured below.

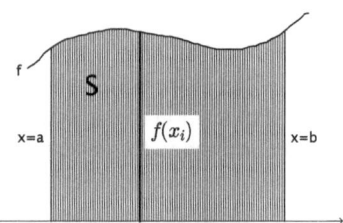

In contrast to thinking about the area of a region in terms of square units, Cavalieri's principle asks us to imagine the region S as "composed" of a continuum of parallel line segments. Conceptualizing area in this way is not particularly helpful for measuring it. Fundamentally, we cannot add up the areas of the line segments to obtain the area of S. (Each line segment has no area and there are infinitely many of them.) But this image of area as an accumulation of line segments, together with the notion of instantaneous rates of change, turns out to be a useful device to take with us into a deeper examination of the proof of FTC.

13.3   Connecting to Real Analysis

The Fundamental Theorem of Calculus is stated in two parts: the first part is about integrating a derivative, and the second part is about differentiating an integral. Although it can be stated for integrable functions, we shall simplify matters slightly and restrict our attention to the smaller class of continuous functions which are more familiar in secondary mathematics.

Theorem (Fundamental Theorem of Calculus for continuous functions)

(i) Let $f : [a, b] \to \mathbb{R}$ be continuous, and $F : [a, b] \to \mathbb{R}$ satisfy $F'(x) = f(x)$ for all $x \in [a, b]$. Then $\int_a^b f = F(b) - F(a)$.
(ii) Let $f : [a, b] \to \mathbb{R}$ be continuous and define $F(x) = \int_a^x f$. Then for any $c \in [a, b]$, $F'(c) = f(c)$.

In the teaching scenario, Ms. Kalili's exercise was designed to illustrate part (ii) of the FTC, and this is the part of the theorem we will give our primary focus. (Several exercises at the end of the chapter ask you to engage with part (i).)

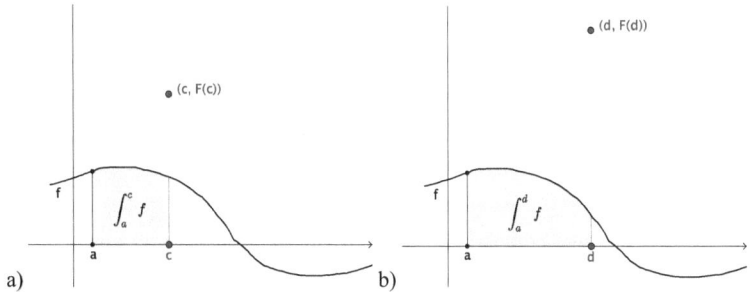

Fig. 13.1 Meaning of F as an area under f, and particular values (**a**) $F(c)$, and (**b**) $F(d)$

Fig. 13.2 Graph of $F(x)$, an accumulation of (positive and negative) area as x increases

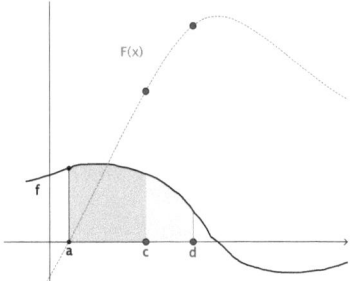

13.3.1 Instantaneous Rates of Area and $F'(c)$ in FTC(ii)

To get an understanding of why FTC(ii) is a valid statement, it helps to have a tangible example for the function F that illustrates the way in which it depends on f. Note that a is fixed; and the variable x is the upper limit of integration so that we can think of $F(x) = \int_a^x f$ as the area under f from a up to x. If we take $x = a$ then $F(a) = \int_a^a f = 0$ since there is no area. Figure 13.1 provides a visual illustration for $F(c) = \int_a^c f$ and $F(d) = \int_a^d f$ where c and d are two x-values in the interval $[a, b]$. The area of the shaded regions gives the values for F, which are also plotted as points in the figure. Notice that $F(d) > F(c)$ in this example because the corresponding area under f is clearly larger.

In the theory of the integral, area below the x-axis comes out negative. What $F(x)$ really measures is the *net* area between f and the x-axis where area that is below the axis is subtracted from the total. This means that when f is positive the total increases, and when f drops below the x-axis this total starts to decrease. The sketch of F in Fig. 13.2 illustrates how $F(x)$ keeps a running account, from a up to x, of the accumulated signed area between f and the x-axis. (Be sure you can make sense of why F begins decreasing in the figure before moving on.)

How does this description of F help us envision F'? For a fixed $c \in [a, b]$ the definition of the derivative tells us $F'(c) = \lim_{x \to c} \frac{F(x) - F(c)}{x - c}$. Although we can relate this to slopes of secant lines on F (as we did in Chap. 9), we will instead consider each part of the expression in terms of f and the associated lengths and

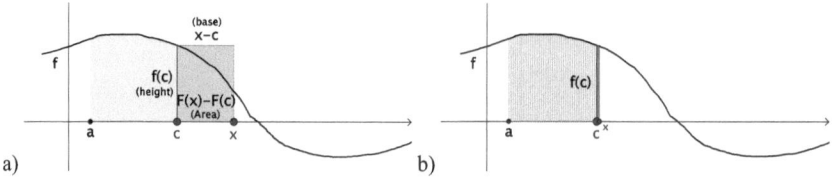

Fig. 13.3 (a) The average rate of change of the area F over the interval $[c, x]$ is reasonably approximated by the height $f(c)$; (b) the instantaneous rate of increase of F at $x = c$ is equal to the height $f(c)$

areas. Taking $x > c$, we can see that, geometrically speaking, $F(x) - F(c)$ is the area under f over the interval $[c, x]$. The quotient $\frac{F(x) - F(c)}{x - c}$ then represents the average rate of change of the area over this interval. Our job is to find the instantaneous rate of change at c.

Figure 13.3 illustrates what happens when we approximate the area $F(x) - F(c)$ with the rectangle of height $f(c)$ and base $x - c$. Because the two regions have the same base, replacing $F(x) - F(c)$ with an actual rectangle lets us approximate the quotient as

$$\frac{F(x) - F(c)}{x - c} \approx \frac{A_{rect}}{x - c} = \frac{f(c) \cdot (x - c)}{x - c} = f(c).$$

How reasonable is this approximation? The image in Fig. 13.3a shows the two areas can be quite different (there, $f(c)(x - c)$ overestimates $F(x) - F(c)$), but Fig. 13.3b illustrates how this approximation improves as x gets closer to c. This observation should be familiar from the theory of the Riemann integral. For continuous functions, the rectangle with height $f(c)$ becomes an increasingly effective estimate for the area under the curve as the interval $[c, x]$ gets smaller. The key is that dividing the area of a rectangle by its base yields its height, and so the limit of the quotient $\frac{F(x) - F(c)}{x - c}$ (which is akin to dividing a rectangular area by its base) converges to $f(c)$ as $x \to c$.

The conclusion, that $F'(c) = \lim_{x \to c} \frac{F(x) - F(c)}{x - c} = f(c)$, can be interpreted as a statement about the instantaneous rate of change. Specifically, it says that the rate of change of the total signed area between f and the x-axis at a point c is equal to $f(c)$. Recall that Cavalieri's principle suggests a mental picture of the region between f and the x-axis as a row of vertical line segments with heights equal to the value of the function. With this image in mind, the instantaneous rate of change for the accumulated area under a function at c should indeed be the height of the function at that point—like imagining the region accumulating its area by incrementally tacking on line segments! For a visual confirmation of this conclusion, you might also note how the slopes on the graph of F in Fig. 13.2 appear to be everywhere equal to the values of f.

13.3.2 Proof of FTC(ii)

Having created a conceptual framework for understanding the statement in FTC(ii), we now offer a proper proof that essentially follows the same line of reasoning, only this time in the language of real analysis.

Proof Recall that we are assuming f is continuous on $[a, b]$ and $F(x) = \int_a^x f$. Given $c \in [a, b]$, our job is to prove $F'(c) = f(c)$.

The definition of the derivative tells us that

$$F'(c) = \lim_{x \to c} \frac{F(x) - F(c)}{x - c}.$$

In the previous discussion, we interpreted $F(x) - F(c)$ to be the area under f on the interval $[c, x]$. This time we write $F(x) - F(c) = \int_a^x f - \int_a^c f = \int_c^x f$ using the additive property of integrals (see Abbott's [1] Theorem 7.4.1). Our task then becomes to show

$$F'(c) = \lim_{x \to c} \frac{\int_c^x f}{x - c} = f(c).$$

We revert to the $\varepsilon - \delta$ definition of functional limit. So letting $\varepsilon > 0$, we must find a δ-neighborhood around c in which $\left| \frac{\int_c^x f}{x - c} - f(c) \right| < \varepsilon$ for $x \neq c$ in this neighborhood.

Because f is continuous at $c \in [a, b]$, we know there is a $\delta > 0$ such that

$$|f(x) - f(c)| < \varepsilon \text{ whenever } |x - c| < \delta.$$

This means $f(c) - \varepsilon < f(x) < f(c) + \varepsilon$ for all x in this δ-neighborhood of c. To simplify the notation, let's focus on the case where $x > c$. (The case $x < c$ is similar.) Because $f(c) - \varepsilon$ and $f(c) + \varepsilon$ serve as lower and upper bounds for f on $[c, x]$, the integral $\int_c^x f$ must be bounded by the areas of lower and upper rectangles (by Abbott's Theorem 7.4.2). Specifically,

$$(f(c) - \varepsilon)(x - c) < \int_c^x f < (f(c) + \varepsilon)(x - c).$$

Dividing this inequality by $(x - c)$ we get

$$f(c) - \varepsilon < \frac{\int_c^x f}{x - c} < f(c) + \varepsilon.$$

In our earlier informal argument, we approximated $\int_c^x f$ with the rectangular area $f(c)(x - c)$ which left us with $f(c)$ after dividing by the length of the base. In the formal proof we have shown that the quotient $\frac{\int_c^x f}{x-c}$ is bounded between $f(c) - \varepsilon$ and $f(c) + \varepsilon$ for all x in our δ-neighborhood, which is precisely what is required to conclude $F'(c) = \lim_{x \to c} \frac{\int_c^x f}{x-c} = f(c)$. $\qquad\qquad\qquad$ $\square$

13.4   Connecting to Secondary Teaching

In the initial teaching situation, the function Ms. Kalili used to motivate the Fundamental Theorem of Calculus represents a positive instance of TP.2; the linear function is a useful special case because of how easy it is to visualize and to work with algebraically. This allowed her students to compute a formula for the integral using geometric means and then realize that their result was an anti-derivative. The teacher's example enabled the students to experiment with the theorem's claim procedurally and computationally, but it did not provide any insight as to why the theorem is true. This falls short of TP.5's emphasis on conceptual understanding, but conceptual understanding is an ambitious goal when it comes to FTC. Taking our time going through the proof—first informally and then formally—illustrates the depth of the mathematical ideas involved. Focusing on instantaneous rates of change and the intuition of Cavalieri's principle are useful ways to glean some clarity and insight out of the formal proof.

13.4.1 Conceptual Explanation

Consider the following continuation of the pedagogical situation:

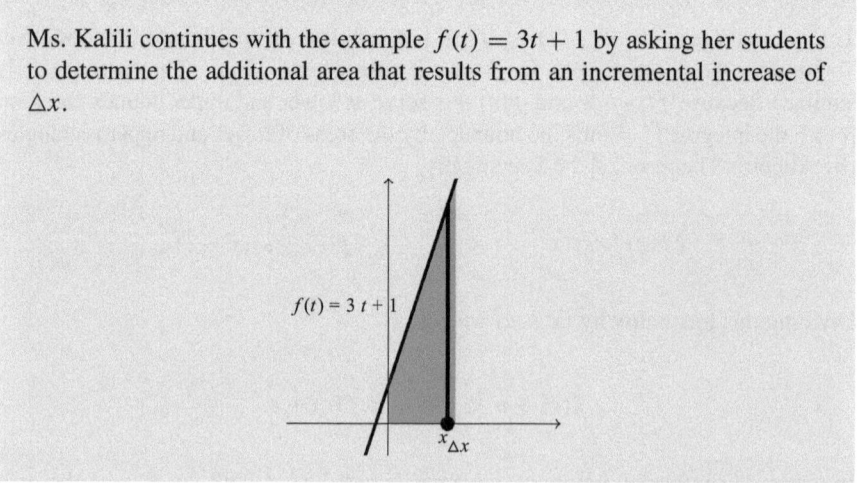

Ms. Kalili continues with the example $f(t) = 3t + 1$ by asking her students to determine the additional area that results from an incremental increase of Δx.

$f(t) = 3\,t + 1$

$x_{\Delta x}$

(continued)

Picking up on Destiny's observation, Ms. Kalili points out the additional area is trapezoidal. She asks students to compute this area:

Destiny: So for that trapezoid, the bases are $f(x) = 3x + 1$ and $f(x + \Delta x) = 3(x + \Delta x) + 1$, and the height is Δx. So the new area is:

$$\frac{1}{2}\Delta x \cdot (3x + 1 + 3(x + \Delta x) + 1)) = \Delta x \cdot \left(3x + 1 + \frac{3}{2}\Delta x\right)$$

Ms. Kalili: Excellent! But instead of a total, let's think about this change in area as a slope or a rate: how much does the area increase per unit increase in x?
Destiny: Oh, so you mean:

$$\frac{\Delta \text{Area}}{\Delta x} = \frac{\Delta x \cdot \left(3x + 1 + \frac{3}{2}\Delta x\right)}{\Delta x} = 3x + 1 + \frac{3}{2}\Delta x$$

Ms. Kalili: Precisely. This is the *rate* at which area increases. But if we want the *instantaneous* rate, we are interested in this rate when $\Delta x \to 0$. This simplifies the expression to

$$3x + 1$$

which represents the instantaneous rate at which the area under this function f increases. Note that this expression—which is the derivative of the integral of the function—is $f(x)$, the height of the vertical segment at x!

In this scenario, the teacher has offered students an opportunity to explore why the derivative of the integral of a function is the value of the function at that point. The additional questions she poses complement the earlier ones by providing insight into what the instantaneous rate of change of an integral should mean in terms of area, and why that instantaneous rate is $f(x)$. This conclusion jives with the intuition from our work with Cavalieri's principle where an incremental increase in x should increase the total area at a rate corresponding to a line segment—namely, one equal to the height of the graph at x. This doesn't mean the area increases by $f(x)$ units— it means that the instantaneous rate the area increases is equal to $f(x)$ units^2 per unit increase in x.

13.4.2 Area and Conceptualizing Instantaneous Rates of Change

Once they become familiar with the machinations of calculus, students might notice something interesting about the familiar formulas for area and circumference of a circle:

$$A_{circ} = \pi r^2 \quad \text{and} \quad C_{circ} = 2\pi r.$$

The circumference is the derivative of the area! Why should this be the case?

In Cavalieri's principle, viewing a region as a continuum of parallel line segments provided useful intuition for making sense of how the Fundamental Theorem of Calculus treats the area. With some modifications, the same type of intuition can be applied to the circle.

Suppose we have a circle of radius r, and we consider what happens to the area when the radius increases by Δr, depicted below.

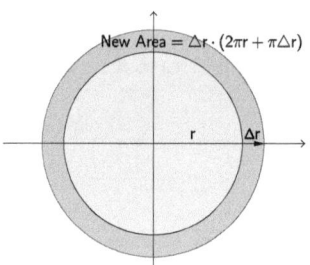

The additional ringed area is:

$$\pi(r + \Delta r)^2 - \pi r^2 = \pi r^2 + 2\pi r \Delta r + \pi(\Delta r)^2 - \pi r^2 = \Delta r \cdot (2\pi r + \pi \Delta r).$$

So the average rate at which area is changing over the interval from r to $r + \Delta r$ is

$$\frac{\Delta r \cdot (2\pi r + \pi \Delta r)}{\Delta r} = 2\pi r + \pi \Delta r.$$

To find the instantaneous rate we take the limit as $\Delta r \to 0$ which yields $2\pi r$—the circumference! Indeed, if we imagine a very small value for Δr in the diagram above, the additional new ring of area would be visually indistinguishable from a circle with circumference $2\pi r$.

The way area increases in a circle as the radius increases is akin to an accumulation of incrementally larger circumferences, providing the intuitive idea that a circle can be viewed as "composed" of a continuum of concentric rings. In the same way that FTC(ii) identifies a line segment, $f(c)$, as the instantaneous rate of change of the area under a function $f(x)$ as x increases (mirroring Cavalieri's principle), using r as the variable in our geometric context produces an accumulation of concentric rings, which explains why circumference is the derivative of area.

But what happens to this picture if we switch our focus to the diameter? First, let's recast the area formula in terms of d:

$$A_{circ} = \pi \left(\frac{d}{2}\right)^2 = \frac{\pi}{4}d^2.$$

Differentiating this expression with respect to d gives

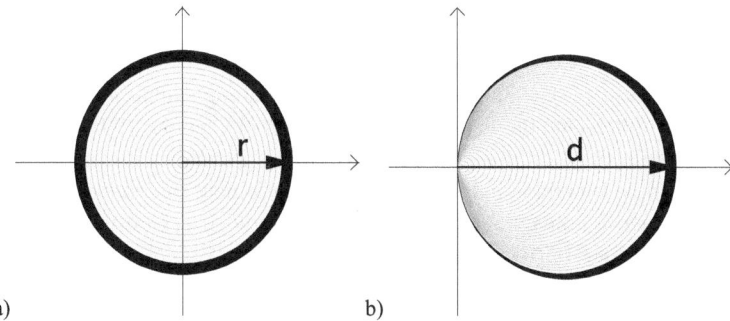

Fig. 13.4 Instantaneous rate of change of a circle's area, based on its (**a**) radius, and (**b**) diameter

$$A'_{circ} = \frac{\pi}{2}d = \frac{1}{2}(\pi d).$$

The conclusion appears to be, as the diameter increases, the instantaneous rate at which area increases is now *half* of its circumference. Visual comparison with the radius example above helps explain why. Increasing the radius by Δr added a uniformly thick ring because the radius is measured from the center. But a diameter is measured from end to end. Incremental changes in d can be considered by aligning the horizontal axis with a *diameter* of a circle. In this case, we fix one end of the circle at $(0, 0)$ and let the horizontal axis represent the circle's diameter. Figure 13.4 gives a visual sense for the rate at which area is accumulating in each of these cases. As we discussed before, increasing the radius (Fig. 13.4a) results in an accumulation of uniformly thick rings—rings that would be indistinguishable from a circumference when that increase is very small. However, incremental increases in diameter only occur on one end, expanding the circle in one direction rather than out from the center. This means the additions to the area are *crescent-shaped* rather than uniformly thick rings—see Fig. 13.4b. It is this detail that accounts for the factor of $\frac{1}{2}$: the crescent-shape is essentially half of a uniformly thick ring, which, with very small increases, would be akin to half a circumference. (You might think about superimposing the crescent-shape in Fig. 13.4b onto the ring in Fig. 13.4a; it is kind of similar to how the area of a triangle is half the area of a parallelogram.) This means increases in d correspond to the circle's area accumulating at an instantaneous rate of half its circumference.[1] In these geometric contexts, the tools of calculus provide us with some new intuitions and ways to think about area and its accumulation.

---

[1] Another way to think about this is that an increase in diameter would mean the radius increasing by half, and so a change in diameter would correspond to half the incremental increase in area for a change in radius.

Problems

13.1 The function $f(t) = 3t + 1$ was used in the teaching situation in this chapter to help students understand the Fundamental Theorem of Calculus (FTC). Think about another example function that you could use to help students understand FTC(ii). Give the example function, and how you would anticipate students reasoning about it. Then, provide a conceptual justification you could use with this example that would explain why FTC(ii) is true.

13.2 The area of a square is often given by $A_{sqr}(s) = s^2$, where s is the side length of the square. Using this formula, provide an explanation for why the derivative, $A'(s)$, is not the perimeter of the square but only *half the perimeter* of the square. Describe which one of the two images from Fig. 13.4—the radius r or diameter d— is most similar in terms of both the variable that is increasing, and the conclusion about the relationship between area and perimeter.

13.3 Consider the depiction of a square defined by length r (below). (i) Write a formula that expresses A_{sqr} in terms of the length r. (ii) Show that $A'(r) = 8r$, which is in fact the perimeter of the square. (iii) Provide an explanation for why the derivative of the square's area, in terms of r, now gives the *full* perimeter of the square, which was not the case in Problem 13.2.

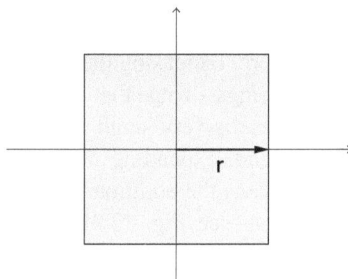

13.4 Our special case of the Fundamental Theorem of Calculus, part (i), states: "Let $f : [a, b] \to \mathbb{R}$ be continuous, and $F : [a, b] \to \mathbb{R}$ satisfy $F'(x) = f(x)$ for all $x \in [a, b]$. Then $\int_a^b f(x)dx = F(b) - F(a)$." Since f is an instantaneous rate of change of F, i.e., $f = \frac{\Delta y}{\Delta x}$, then $f \Delta x = \Delta y$. And since an integral is an accumulation, one way to think about $\int_a^b f(x)dx$ would be an accumulation of these Δys. These Δys are, in fact, line segments; ones that represent vertical changes of F. As Δx tends to 0, these Δys become increasingly less like segments, and more like points. Use this idea to provide a conceptual explanation for why the observation of the FTC(i)—that this accumulation amounts to $F(b) - F(a)$—would be true.

13.5 The function $f(t) = 3t + 1$ was used in the teaching situation in this chapter to help students understand FTC(ii). Describe how you could use that example—or pick another one—to illustrate FTC(i).

13.6 Look at a proof of the Fundamental Theorem of Calculus, part (i) (cf., Abbott's Theorem 7.5.1(i)). Relate your conceptual explanation from Problem 13.4 to the proof by explaining the big ideas of the proof and how they relate to your explanation.

13.7 Consider again the two parts of the FTC stated in this chapter:

(i) Let $f : [a, b] \rightarrow \mathbb{R}$ be continuous, and $F : [a, b] \rightarrow \mathbb{R}$ satisfy $F'(x) = f(x)$ for all $x \in [a, b]$. Then $\int_a^b f = F(b) - F(a)$.

(ii) Let $f : [a, b] \rightarrow \mathbb{R}$ be continuous, and $F(x) = \int_a^x f$. Then for any $c \in [a, b]$, $F'(c) = f(c)$.

In part (i) the choice of a and b seems to be critically important in terms of giving the final value of the area under the function f ($\int_a^b f = F(b) - F(a)$). Yet in part (ii), a and b are "just endpoints" and do not really matter much at all. Calculus students commonly ask why the particular value of a does not matter, since it appears as part of the definition of $F(x)$ (i.e., $\int_a^x f$). Using ideas from this chapter, first write a justification for why the choice of a 'does not matter' in part (ii) so long as the function is still continuous on the chosen interval. You might consider creating a new, arbitrary, constant k such that $a \le k \le c$. Then, write an explanation emphasizing the important ideas of the justification that would be good to give to a high school senior.

13.8 Consider the general statement of FTC(ii), not the special case discussed in this chapter. Abbott states the theorem as follows (Theorem 7.5.1 (ii)):

(ii) Let $g : [a, b] \rightarrow \mathbb{R}$ be integrable, and for $x \in [a, b]$, define $G(x) = \int_a^x g$. Then G is continuous on $[a, b]$. If g is continuous at some point $c \in [a, b]$, then G is differentiable at c and $G'(c) = g(c)$.

Notice that g in this statement is integrable (whereas we used a stricter condition that it be continuous). TP.1 suggests that one be aware of assumptions, which in this case would be the specific conditions under which the theorem and its conclusions are valid. First, specify the additional condition (besides g being integrable) that is necessary for the conclusion, $G'(c) = g(c)$, to be valid. Then, look at one or two discontinuous functions g, and their points of discontinuity at $x = c$, and explain why the conditions are necessary to warrant the conclusion.

References

1. Abbott, S. (2015). *Understanding analysis* (2nd ed.). New York, NY: Springer.

Afterword

In the Preface we introduced the primary goal of this text: to consider how ideas and ways of thinking that you engage with in real analysis might be valuable in teaching secondary mathematics. We focused on two distinctive kinds of crossover: content connections and process connections. Content connections are those where secondary mathematics is generalized in real analysis or those where real analysis provides a specific example of a general idea in secondary mathematics. A process connection highlights a mathematical procedure or approach that is important in both secondary mathematics and real analysis. In Chap. 1 we outlined a set of six teaching principles (TPs). These principles—which have their roots in both mathematical practice and pedagogical practice—became the foundation for how we discussed secondary teaching situations throughout the text. In this afterword, we summarize the connections we explored in each of the chapters and discuss them in relation to the two types of connections and the six teaching principles.

In Chap. 2, the teaching situation showed a student converting an addition task from fractions to decimals and claiming that fractions have only one decimal representation. We connected that to real analysis by exploring the claim that decimal representations are unique, as an example of a more general inquiry about "When are two numbers equal?" What we found was that every terminating decimal has exactly two different infinite decimal representations; we showed they were equal by ruling out the possibility that the difference between them was a positive number, $\varepsilon > 0$. There are a few things that you might bring to your classrooms from this chapter. First, decimals do not always provide a unique representation of a real number! Second, you can and should ask students to think about the ideas, not just the procedures. Third, you can do this either by asking follow-up questions that request further explanation or posing new problems that force students to confront incorrect assumptions in their claims (TP.2).

In Chap. 3, the teaching situation was about teaching students to approximate irrational numbers—specifically, a standard from the Common Core State Standards for Mathematics (CCSSM) in the U.S. [1]. We concluded that such processes should be iterative and consistent and that their error-bounds need to become arbitrarily small in order to be certain we could obtain an approximation that would be

© The Author(s), under exclusive license to Springer Nature Switzerland AG 2022
N. H. Wasserman et al., *Understanding Analysis and its Connections to Secondary Mathematics Teaching*, Springer Texts in Education,
https://doi.org/10.1007/978-3-030-89198-5

close to the desired number. This CCSSM standard has a direct analog in real analysis—being able to characterize when, and show that, a sequence converges. What we see from studying the real analysis content is justification for why the secondary approaches actually work, as well as more general principles for what approximation processes should entail. In this case, the real analysis is more general than the K-12 mathematics in that it provides formal definitions and criterion for sequences and their convergence. Approximating irrational numbers is an instance of this in secondary mathematics. The takeaway is that real analysis can provide important insights about general processes used in secondary classrooms, and that this knowledge can allow teachers to be more flexible in their classrooms, opening opportunities to give students more agency to explore. Doing so puts a spotlight on TP.6, emphasizing multiple approaches and multiple explanations.

In Chap. 4, the teaching situation showed a student rounding in the middle of a problem rather than at the end. The student's answer was very near to the answer he would have gotten by rounding at the end. We explored the question, "Why is the rule to round at the end?" We used more general ideas and proofs from real analysis about sequence convergence with sums and products as a tool to give precise quantification for how error can accumulate when we perform arithmetic operations on rounded values. There are a few things you might take back to your classroom. First, that the ideas from real analysis allow us to provide an explanation for the rule, "round at the end, not in the middle," that goes beyond "it might cause problems." This is in accord with TP.5, as the insights allow for knowing how much error might be introduced as a result of different steps. Perhaps more importantly, teachers could use these ideas to design problems for students to solve that showcase that the error introduced by rounding can become quite large (TP.2).

In Chap. 5, we focused on logical statements in mathematics. This is an example of a process connection. In both the teaching situation and in the real analysis we had examples of statements that had logical relations that were difficult to parse, we translated them into more formal versions, and we explored several confusions that might emerge as the intended relationships get expressed and interpreted. Specifically, the 'iff', 'is', and 'converse' confusions. We should take away that we will need to support our students in learning the logical and semantic culture of mathematics (TP.3). For example, by asking students whether a statement is meant as a definition or a property; or is about a criterion for or a description of a concept. In addition, as teachers, it is good to try to be less ambiguous with our statements, but also important to work on grammatically varying the way in which we describe relationships in order to help convey to students the intended nuances. Finally, in our responding to students, we should try to explore their thinking—are they misinterpreting the content, or the logic? To help them learn, we need to respond to, and ask questions about, their actual thinking and not just our interpretation of it.

In Chap. 6, we made another process connection, this time exploring how our choices of definition matter in mathematics. We did so using the secondary content of trapezoids, which have two logically different definitions, and the real analysis content of continuity, where some of our definitions were competing and others equivalent. Some ideas we might take away include conveying that real people write definitions, typically to try to capture their "intuitive" sense of the object. And that the choice of definition matters in quite a few ways, including how subsequent ideas (like isosceles trapezoids) get defined, as well as what are examples and non-examples since mathematical definitions are stipulative. In mathematics, if you have two logically equivalent choices for a definition, you might pick one to serve as the definition based on it making some particular proof a bit easier; but when definitions are not logically equivalent, you might think about whether one choice makes the development of subsequent theorems more natural. Regardless, it is important to help students understand both the role that definitions play in mathematics (TP.2), as well as the human element involved in the process of defining.

In Chap. 7, we made another process connection. In both the teaching situation and the real analysis we looked for and identified implicit assumptions in mathematical statements. This process in the real analysis context helped us unpack, from the explicit statement of the theorem, some of the potentially implicit assumptions of the Intermediate Value Theorem. Notably, being both continuous and defined on an interval were important aspects of the theorem's conditions—ones which also became relevant in the theorem's proof. This attention to unpacking seemingly small details also helped us think further about the teaching situation. Namely, the process of learning means that students often do not know to make certain assumptions explicit. Teaching should respond in ways that both acknowledge the contributions students make, as well as draw out what has been assumed (TP.1). We should take away the idea that it is useful to engage in this practice in our own doing of mathematics, and in the ways we present material in class and how we listen to and respond to students.

In Chap. 8, we explored how solving equations in secondary mathematics makes use of inverse functions, and the potential problems that arise. Studying the real analysis content led us to some new ways to characterize when continuous functions have inverses, which also had implications for how we might restrict the domain of functions in order for them to be invertible. That is, the real analysis content helped to answer a more general question, but one which has many examples in secondary mathematics. There are a few ideas we should take away from this chapter. First, we now have an explanation for the "horizontal line test." And when a function does not have an inverse, we now know how to restrict the domain to create a partial inverse; namely, choose a domain on which the continuous function is strictly increasing (or decreasing). Second, these domain restrictions play an important role when inverse functions get applied to the equation solving process. Specifically, applying an inverse function allows us to find a unique solution, often on a restricted domain; properties such as symmetry and periodicity allow us to recoup other solutions that might have been lost in this process, and analyzing the different domains allows us to recognize when we might have to remove solutions that have been accidentally gained. All of these speak to the importance of TP.5 in the classroom. We always want to be able to explain the ideas behind a rule.

In Chap. 9 we explored questions about slopes and derivatives, where the real analysis content helped us to explain a rule about discontinuous functions and their derivatives. The real analysis in this chapter provided a more general tool to think about a specific claim from calculus. It involved recognizing that the slope computation in the functional limit definition could be seen as either a series of procedures on several constituent parts, or as a singular object—a (new) function. The ideas to take away from this chapter are that we developed a new tool, the Secant Slope Function, which we used to show and explain why a function with a "jump discontinuity" would not have a defined derivative at that jump. (The explanation can be adapted for other types of discontinuities as well!) Notably, in contrast to a typical proof of this fact, this approach allowed for a more insightful rationale for why this would be true in the first place (TP.5).

In Chap. 10, we explored a process connection between secondary mathematics and real analysis. In both contexts, it is important to ask, "What is the scope of this statement (or proof)?" Such attention to scope provides an avenue to interrogate potentially hidden assumptions in our—and our students'—claims (TP.1). And by analyzing precisely what ideas a justification relies on, we can also determine whether a statement can be expanded or should be contracted. For example, if a valid proof does not use a particular hypothesis, eliminating it broadens the scope of the original claim. We also saw that building the most general version of a claim (in that chapter, the power rule for derivatives) may require a number of sequential and cumulative steps. A take away for the secondary classroom is about "rules that expire." While it is true in third grade that students cannot take the square root of a negative number, that rule does not continue to hold up in later grades. Teaching is filled with situations like this. As two possible paths forward, we discussed making more general statements, or sharpening and foreshadowing the limitations. If you want to read about some teachers who did this in their secondary classes after a real analysis class like yours, read Wasserman et al. [2]!

In Chap. 11, we again explored a process connection, in which we looked at modeling the complex with the simple in both secondary mathematics and real analysis. In relation to the classroom, we saw that students might benefit from such approaches, by looking at the example of modeling circles with regular polygons and at the conclusions we were able to deduce about circles from this approach (TP.4). Real analysis offered us the opportunity to practice this skill using polynomials, which we understand well, to model a function that appears to be nice (the "bell curve") but is deceptively difficult. A few ideas to take away from this chapter include that this modeling process can provide us with specific conclusions about lesser known objects, often by giving us bounds for how they would have to behave in relation to what we know about other objects. This practice of modeling more complex ideas with simpler ones is similar to Polya's notion of solving a related, easier problem before solving the harder one you began with. When thinking about yourself or your students, starting from familiar contexts is helpful!

In Chap. 12, we explored the notion of area-preserving transformations. In the teaching situation, we explored Cavalieri's theorem and how we can use "segment-skewing" to transform geometric shapes while maintaining their area. In real analysis, we connected this idea to the Riemann integral approach of determining the area under a curve by using increasingly thin rectangles. This is a case where the real analysis content is a more specific version of the general idea from K-12 mathematics! Two ideas to take away from this section are, first, the meaning and use of Cavalieri's theorem (which is part of the CCSSM standards)—not just with volumes but with areas. Second, having multiple explanations, and not just one, can help expand the opportunities students have to reason about mathematical ideas (TP.6).

In Chap. 13 we focused on understanding the Fundamental Theorem of Calculus (FTC) and, in doing so, distinguished (again) between stating a process and explaining or justifying that process. Understanding average and instantaneous rate of change in the context of area, as well as Cavalieri's approach to conceptualizing area, helped make sense of why the seemingly unrelated processes of finding an anti-derivative and computing area under a curve are, in fact, intertwined. While you should definitely take away at least one way to justify why the FTC is true, we also want the distinction between stating a process and explaining to be clear. An explanation needs to give a reason why something is true in a way that students can understand (TP.5). This explanation could come in the form of a proof, or maybe an example that illustrates the critical ideas, but fundamentally it needs to answer the question, "Why does my process accomplish what I want it to?"

In each chapter, we have explored a different connection between secondary mathematics teaching and real analysis. This is not an exhaustive list of either content or process connections between the two domains, only some of the ones we found to be clear and useful. We encourage you to continue to think about the relationships between secondary mathematics and real analysis—or other mathematics courses you take. You might ask yourself how specific mathematical ideas could help you better understand the content of secondary mathematics, or how mathematical processes might be informative for your own classroom teaching. Regardless, continue to be curious about mathematics and your students' thinking about mathematics. A lifetime of learning awaits!

References

1. Common Core State Standards in Mathematics (CCSSM). (2010). Retreived from: http://www.corestandards.org/the-standards/mathematics.
2. Wasserman, N., Weber, K., Fukawa-Connelly, T., & McGuffey, W. (2019). Designing advanced mathematics courses to influence secondary teaching: Fostering mathematics teachers' "attention to scope." *Journal of Mathematics Teacher Education, 22*(4), 379–406.

Index

A

Abbott's
  definition, 82, 89
  text, 53, 54, 62, 134, 147, 175
  theorem, 46, 62, 100, 166, 197, 203
Actual error, 28, 43, 53
Adrian's approach, 42
Algebraic Limit Theorem, 51
  real analysis, 45
    Algebraic Limit Theorem for sequences,
      45–46
    error accumulation, implications for,
      46–48
    visualizing potential error inequality,
      48–49
  secondary mathematics
    approximation and error accumulation,
      43–45
    problematizing teaching and
      pedagogical situation, 42–43
  secondary teaching, 49
    error accumulation, applying principles
      of, 49–51
    peripheral, 51–52
  for sequences, 45–46
  teaching problem, 41–42
Algebraic Limit Theorems and error
      accumulation, 43–51
Approximations, 26–28
  Algebraic Limit Theorems and error
      accumulation, 43–45
  features of, 32–33
  heuristic behind, 35
  as infinite sequences, 29–30
  for rational number, 37
  and sequence convergence, 30–32
Arbitrarily close to zero process, 33, 37
Arbitrary closeness, 30

Archimedean property, 13
Archimedes model, 5, 172, 173
Attention to scope
  derivative statement, 146
  exponent statement, 145
  justification scope, 152–153
  pedagogical practice, 150–152
  perimeter statement, 145
  real analysis course
    definition-theorem-proof model, 146
    power rule, 148–150
    theorems and proofs, 147–148
  teaching and the pedagogical situation,
      144–145
Axiom of Completeness, 13

B

Biconditional statement, 57
Bounding interval, 27, 28, 30

C

Cartesian product, 96
Cavalieri's principle
  areas of ellipses, 186
  areas of polygons, 184–186
  composite transformation, 187–188
  cut-reassemble approach, 180, 184
  FTC, 194
  geometric object, 187
  integral properties, 183–184
  justification, 181–183
Chain rule, 149, 156
Classroom communication, 64, 143
Classroom, navigating disagreements, 136–138
Common Core State Standards for
      Mathematics (CCSSM), 26

Competing definitions, 75
Completeness, 13
Complex numbers, 20
Computations as singular objects, 129–130
Concept function, 93
Conditional statement, 56
Conditions in proofs, 99–101
Consistency, 32
Consistent process, 27, 32, 33
Continuity and definitions
    competing definitions, 75
    equivalent definitions, 74
    real analysis statement
        Abbott's definition, 82, 89
        choosing a definition, 83–84
        continuous functions, 79
        Dirichlet function, 80
    secondary teaching
        definitions and theorems, 86–87
        isosceles trapezoids, 85–86
    teaching and pedagogical situation
        isosceles trapezoid, 77–78
        trapezoids, 76–77
    teaching problem, 73–76
Convergence, 30–32
Converse confusion, 59
Cut-reassemble algorithm, 178, 179, 184

D
Decimal representations, 11–12
Definition-theorem-proof model, 146
Derivative statement, 146
Differentiability requires continuity, 129
Dirichlet function, 4, 23, 80
Divergence criteria and logic in communication
    algebraic solving process, 69
    conditional statement, 59–60
    linear equations, 71
    real analysis statements
        convergence theorems, 60–62
        logical implications about divergence,
            63–64
    secondary teaching
        content and logic matters, 65–66
        grammatical variation, 67–68
        logical issue of the converse, 66–67
    teaching and pedagogical situation
        biconditional statement, 57
        conditional statement, 56
        converse confusion, 59
        expressed relationship, 57
        iff confusion, 58
        intended relationship, 57

        interpreted relationship, 57
        is confusion, 59
        Quad. with Cong. Diag, 59
        Trap. with Cong. Diag, 59
    teaching problem, 55–56
Domain restrictions, strict monotonicity,
    119–120

E
ε-approach
    defining equivalence of real numbers, 14
Equivalence classes, 11–12
Equivalent definitions, 74
Equivalent real numbers
    real analysis, connecting to, 12–13
        ε-approach, defining equivalence of real
            numbers, 14
        real numbers, implications for, 15–16
    secondary mathematics
        equivalence classes and decimal
            representations, 11–12
        problematizing teaching and
            pedagogical situation, 10–11
    secondary teaching, connecting to, 16
        exploring infinite decimals with
            students, 17–18
        number sets, progression of, 18–19
    teaching problem, 9–10
Error-bounds, 27–28, 43
Essential discontinuity, 139
Euclid, 4, 5
Euclid's elements, 55
Exclusive definition, 76–78, 86, 87
Exponent statement, 145, 153
Expressed relationship, 57
Extraneous solutions, 122–124

F
Function definition, 95–96
Fundamental Theorem of Calculus (FTC)
    Abbott theorem, 203
    Cavalieri's principle, 194
    conceptual explanation, 198–199
    conceptual justification, 202
    continuous functions, 194
    instantaneous rates of area, 195–196
    instantaneous rates of change, 199–201
    proof of, 197–198
    rate of change, 193
    teaching and pedagogical situation, 193
    teaching problem, 191–192

H
Heuristic behind approximations, 35

I
iff confusion, 58
Implicit assumptions, IVT
    in classroom, 102–103
    in secondary mathematics, 103–104
Inclusive definition, 76–78, 85, 87
Inconsistent method, 27
Infinite decimal representations, 11
Infinite decimals
    real analysis, connecting to, 12–13
        ε-approach, defining equivalence of real
            numbers, 14
        real numbers, implications for, 15–16
    secondary mathematics
        equivalence classes and decimal
            representations, 11–12
        problematizing teaching and
            pedagogical situation, 10–11
    secondary teaching, connecting to, 16
        exploring infinite decimals with
            students, 17–18
        number sets, progression of, 18–19
        with students, 17–18
        teaching problem, 9–10
Infinite series, 54
Integers, 20
Intended relationship, 57
Intermediate Value Property (IVP), 37, 107
Intermediate Value Theorem (IVT)
    function, 95–96
    geometry test, 105
    implicit assumptions
        in classroom, 102–103
        in secondary mathematics, 103–104
    real analysis statement
        conditions in proofs, 99–101
        continuous function, 99, 107
        discontinuous function, 98
    teaching and the pedagogical situation,
        94–95
    teaching problem, 93–94
Intermediate Zero Theorem, 97
Interpreted relationship, 57
Inverse function, 115–116
    conventional domain, 124
    different function, 126
    extraneous solutions, 122–124
    missing solutions, 120–122
    solve trigonometric equations, 113–115
    and strictly monotonic, 117–118

Irrational decimal approximations, sequence
        convergence and
    real analysis, 28–29, 38
        approximation processes and sequence
            convergence, 30–32
        features of approximation processes,
            32–33
        infinite sequences, approximation
            processes as, 29–30
    secondary mathematics
        approximations and error-bounds,
            27–28
        problematizing teaching and
            pedagogical situation, 26–27
    secondary teaching, 33
        general heuristic behind
            approximations, 35
        student-centered instruction, 33–34
    teaching problem, 25–26
Irrational number, 26, 32, 35
is confusion, 59
Iterative process, 29

J
Jump discontinuity, 139

L
Location Theorem for Polynomials, 38
Lower bound, 27

M
Mathematics, 4
Missing solutions, 120–122
Multiple theories of learning, 4

N
Natural numbers, 9
Navigating disagreements in classroom,
        136–138
Nested Interval Property, 13
Non-terminating decimal, 16

O
oblique parallelogram, 178, 179
One-to-one to strict monotonicity, 119

P
Pedagogical mathematical practices, 2
Pedagogy, 1

Perimeter statement, 145
Piece-wise defined function, 140
Polynomials, 5, 38
Potential error, 28, 43, 52
Probability statement, 155
Product, accumulation of, 47
Pythagorean Theorem, 6, 42

Q
Quotient rule, 47, 48, 50

R
Random real number, 37
Rational number, 5, 9, 11–13, 16, 17, 20
Real number, implications for, 15–16
Reciprocal, accumulation of, 47
Removable discontinuity, 139
Riemann-integrability of functions, 177–190
Rounding, 41, 42

S
Scalar product, accumulation of, 46
Scalar product rule, 47
Secant slope function
    computations, 134
    definition, 132
    derivative as function, 133–134
    derivatives and continuity, 134–136
    essential discontinuity, 139
    jump discontinuity, 139
    real analysis statement, 130–133
    removable discontinuity, 139
    teaching problem, 127–128
Secondary mathematics
    Algebraic Limit Theorems and error
        accumulation
        approximation and error accumulation,
            43–45
        problematizing teaching and the
            pedagogical situation, 42–43
    equivalent real numbers
        equivalence classes and decimal
            representations, 11–12
        problematizing teaching and
            pedagogical situation, 10–11
    sequence convergence and irrational
        decimal approximations
        approximations and error-bounds,
            27–28
        problematizing teaching and
            pedagogical situation, 26–27

Secondary teaching
    Algebraic Limit Theorems and error
        accumulation, 49
        error accumulation, applying principles
            of, 49–51
        peripheral, 51–52
    equivalent real numbers, 16
        exploring infinite decimals with
            students, 17–18
        number sets, progression of, 18–19
    sequence convergence and irrational
        decimal approximations, 33
        general heuristic behind
            approximations, 35
        student-centered instruction, 33–34
Segment-skewing transformations, 181, 184,
    187, 189, 190
Sequence convergence
    approximation processes and, 30–32
    real analysis, 28–29, 38
        approximation processes and sequence
            convergence, 30–32
        features of approximation processes,
            32–33
        infinite sequences, approximation
            processes as, 29–30
    secondary mathematics
        approximations and error-bounds,
            27–28
        problematizing teaching and
            pedagogical situation, 26–27
    secondary teaching, 33
        general heuristic behind
            approximations, 35
        student-centered instruction, 33–34
        teaching problem, statement of, 25–26
Sequences, 29–30
Solving equations, 118–119
Solving trigonometric equations, 113–115,
    122, 155
Stipulative definition, 73
Strict monotonicity
    domain restrictions, 119–120
    inverse function and, 115–116
    one-to-one, 119
Student-centered instruction, 33–34
Sum, accumulation of, 47

T
Taylor polynomials
    Abbott's text, 175
    converges pointwise, 167
    converges uniformly, 168

π with regular n-Gons, 161–164
hands-on activity, 160
more with circles, 171–172
pedagogical practice, 170–171
reason for analysis, 169–170
standard normal distribution, 165
Taylor coefficients, 165
Taylor's formula derivation, 166–167, 174
Teaching Principles (TPs), 1
    acknowledge and revisit
        assumptions andmathematical
        constraints/limitations, 3
    mathematical explanations, avoid giving
        ruleswithout accompanying, 5–6
    multiple explanations, 6–7
    test and illustrate mathematical ideas, 3–4
    theoretical orientation of, 1–2
    TP.1, 2, 3, 10, 16, 17, 21, 42, 71, 94, 95,
        101, 102, 112, 139, 141–142, 144,
        145, 150, 152, 174, 187, 203, 207,
        208
    TP.2, 2, 3–4, 21, 23, 50, 66, 84, 85, 102,
        141, 175, 190, 192, 198, 205–207

TP.3, 2, 4–5, 38, 39, 64, 107, 112, 175, 206
TP.4, 2, 5, 19, 35, 90, 170, 174, 208
TP.5, 2, 5–6, 26, 33, 43, 49, 51, 113, 119,
    128, 136, 138, 157, 177, 185, 198,
    206–209
TP.6, 2, 6–7, 34, 56, 67, 68, 108, 178, 184,
    185, 206, 209
underpinning mathematical interpretation,
    expose logic as, 4–5
use simpler objects, 5
Terminating decimal, 12, 16
Truncated decimal approximations, 31, 32, 36
Truncating, 53

U
Undoing process, solving equations, 111
Univalence property, 96
Upper bound, 27

V
Vertical line test (VLT), 154